CHEMISTRY RESEARCH AND APPLICATIONS

POLYPROPYLENE

ADVANCES IN RESEARCH AND APPLICATIONS

CHEMISTRY RESEARCH AND APPLICATIONS

Additional books and e-books in this series can be found on Nova's website under the Series tab.

CHEMISTRY RESEARCH AND APPLICATIONS

POLYPROPYLENE

ADVANCES IN RESEARCH AND APPLICATIONS

THÉODORE MARLEAU
EDITOR

nova science publishers
New York

DOI: https://doi.org/10.52305/PKXW6412

Library of Congress Cataloging-in-Publication Data

ISBN: 978-1-68507-378-7

Published by Nova Science Publishers, Inc. † New York

Contents

PREFACE

Polypropylene is a thermoplastic polymer with a wide variety of applications. Researchers are continually finding new opportunities for the use of polypropylene, as discussed in the five chapters of this book. Chapter one aims to show the feasibility of post-consumer polypropylene recycling to obtain 3D printer filaments and presents some necessary characteristics of the materials used in FDM technology to ensure final parts with good properties. Chapter two provides an overview of polypropylene nanocomposites, types of nanofiller, and their applications in the industry. Chapter three explores the potential of polypropylene-based composites as an alternative to conventional radiation shielding materials. Chapter four reviews polymer monoliths' recent environmental and bioanalytical application in several polypropylene tubes, such as pipette tips, syringe barrels, spin columns, and ink-pen tubes. Lastly, chapter five investigates the effects of calcium, aluminum and zinc stearates as pro-oxidant agents on the degradation of polypropylene bags under natural weathering.

Chapter 1 - Polypropylene (PP) materials have been valuable molds and housing to construct low-cost and disposable polymeric monoliths for chromatographic analyses and solid-phase extraction in various matrices exploring reversed-phase, ion exchange, and molecular recognition retention mechanisms. These PP tubes are cheap, chemically stable, and resistant to moderate pressure. In both liquid chromatography and solid-phase extraction, the covalent bonding between the monolith and the PP internal surface prevents channeling and keeps the device working in high performance. The most used approach to this goal is the grafting of pendant vinyl groups to the inner walls of the PP housing by photografting benzophenone to create a surface photoinitiator which mediates the photografting of a thin layer of cross-linked ethylene glycol dimethacrylate (EDMA), thus forming an assortment of terminal vinyl groups useful to anchor the monolith inside the housing. This chapter revises polymer monoliths' recent environmental and bioanalytical applications in several PP tubes, such as pipette tips, syringe barrels, spin columns, and ink-pen tubes.

Chapter 2 - Currently, plastic products have generated severe environmental problems due to the inadequate destination of post-consumer material, which justifies the possibilities of reusing these materials. One of the biggest problems is the large volume that these materials occupy in landfills, and their current low rate of degradation, making their soil complex. Thus, rapid prototyping technology has conquered the market as an innovative technique and, more recently, through using a sustainable source or recycling of polymeric materials, such as polypropylene (PP). Polypropylene is a semi-rigid, lightweight material commonly used in storage and packaging applications. However, the semi-crystalline structure of the material causes great deformation in 3D printed parts on cooling, which is a significant challenge for 3D printing. On the other hand, polypropylene is a strong polymer with excellent fatigue resistance, making it suitable for low-strength applications such as hinges, belts, and leashes. Thus, this chapter aims to show the technical-scientific feasibility of post-consumer polypropylene recycling to obtain 3D printer filaments and present some necessary characteristics of the materials used in FDM technology to ensure final parts with good properties. Furthermore, the use of sustainable polypropylene composite filaments in this area was discussed.

Chapter 3 - This chapter discusses the effects of calcium, aluminum and zinc stearates as pro-oxidant agents on the degradation of polypropylene (PP) bags under natural weathering. Samples were collected periodically and characterized. Tensile properties were evaluated, while the structural modifications were investigated using Fourier transform infrared spectroscopy (FTIR) and melt flow index (MFI). Optical microscopy and scanning electron microscopy (SEM) were carried out. Furthermore, thermal properties were investigated using differential scanning calorimetry (DSC) analysis. The results showed a decrease in elongation and stress at break to very low values, an increase in MFI and carbonyl index over time due to chains' scissions. Changes in melting temperature and crystallinity were also observed after natural weathering. A biodegradation study was carried out for aged formulations during a year. The results showed that biodegradation of the formulations containing pro-oxidants occurred. All the considered properties were affected by the presence of the three pro-oxidant agents used in this study. Furthermore, pro-oxidants can be classified as follows: Zinc stearate > Calcium stearate>Aluminum stearate.

Chapter 4 - Polypropylene (PP) is a thermoplastic polymer with a wide range of uses, from the automotive industry to textile and food packaging. In addition to these known application areas of PP, the use of PP in radiation shielding applications is a virgin area. Although there are currently limited scientific studies on PP-based radiation shielding materials, the commercially advantageous properties of PP such as its low cost, lightweight, easy process, high chemical resistance make PP attractive for using it in the production of radiation shielding materials. From this point of view, the success of composites containing polypropylene in radiation shielding has been reviewed within the scope of the current scientific literature in the first part of the chapter. The second part of the chapter has been devoted to prepare the novel PP-based microstructured composites consisting of natural mineral and ceramic reinforcing particles for radiation shielding. For this purpose, Yahyalı Stone (YS) and the bismuth-based copper oxide layered ceramic micro powders have been used for reinforcing particles for the PP matrix. The radiation shielding performances of the bulk PP/YS and PP/$Bi_2Sr_2Ca_2Cu_3O_{10+x}$ composites with different mass percentages of micro reinforcing particles have been obtained experimentally and theoretically. The radiation shielding abilities of the samples have been evaluated in the context of linear attenuation coefficient (*m*), half-value layer (HVL), mean free path (MFP), effective atomic number (Z_{eff}), and effective electron density (N_{eff}). The experiments have been carried out by experimental setup equipped with NaI(Tl) detector and the Ba-133, Cs-137, and Co-60 point radioactive sources that emit gamma rays with energies ranging from 81 keV to 1332 keV. The theoretical calculations have also been realized by using WinXCom software. As a result, it has been concluded that the highest reinforcing particle contributed PP composites show significant radiation attenuation against X-rays and low energetic gamma-rays. In this context, the mentioned composites can be considered to have the potential to replace conventional radiation-shielding materials such as lead and concrete, which have serious disadvantages in terms of toxicity and weight.

Chapter 5 - In recent years, one of the advances in plastic and composite industries has been the development and commodification of polymer-based nanocomposites. Polymer nanocomposites have recently become the subject of intensive research due to their excellent properties such as electrical and thermal conductivity, toughness and stiffness, mechanical strength, higher barrier to moisture and gases, and superior flame retardancy when compared with the pure polymer or conventional composites (micro-and macro composite). Mechanical properties of the composite and nanocomposite polymers, including compressive modulus, rate-dependent modulus, stress relaxation behaviors, dynamic viscoelastic properties, and fracture mechanics, strongly depend on the homogeneity of the filler dispersion. In this chapter, the authors will provide a brief overview of polypropylene nanocomposites, types of nanofiller, and their applications in the industry. The basic principle of fracture mechanics and crack propagation in polymer materials is described. The effect of different nanoparticles on mechanical properties, morphology, and fracture mechanism of polypropylene nanocomposites are discussed.

In: Polypropylene
Editor: Théodore Marleau
ISBN: 978-1-68507-378-7

Chapter 1

POLYPROPYLENE TUBES EFFICIENTLY HOUSE POLYMER MONOLITHS FOR LIQUID CHROMATOGRAPHY AND SOLID-PHASE EXTRACTION

Fernando H. do Nascimento and Jorge C. Masini*
Instituto de Química, Universidade de São Paulo,
São Paulo, SP, Brazil

ABSTRACT

Polypropylene (PP) materials have been valuable molds and housing to construct low-cost and disposable polymeric monoliths for chromatographic analyses and solid-phase extraction in various matrices exploring reversed-phase, ion exchange, and molecular recognition retention mechanisms. These PP tubes are cheap, chemically stable, and resistant to moderate pressure. In both liquid chromatography and solid-phase extraction, the covalent bonding between the monolith and the PP internal surface prevents channeling and keeps the device working in high performance. The most used

* Corresponding Author's E-mail: jcmasini@iq.usp.br.

approach to this goal is the grafting of pendant vinyl groups to the inner walls of the PP housing by photografting benzophenone to create a surface photoinitiator which mediates the photografting of a thin layer of cross-linked ethylene glycol dimethacrylate (EDMA), thus forming an assortment of terminal vinyl groups useful to anchor the monolith inside the housing. This chapter revises polymer monoliths' recent environmental and bioanalytical applications in several PP tubes, such as pipette tips, syringe barrels, spin columns, and ink-pen tubes.

Keywords: proteins, biological fluids, toxic metals, emerging pollutants, environmental samples

INTRODUCTION

Monolithic stationary phases in liquid chromatography (LC) emerged in the early 1990s (Guiochon 2007; Svec and Huber 2006), and the first report on their use for solid-phase extraction (SPE) appeared in 1998 (Xie, Svec, and Fréchet 1998). These materials gained popularity because they are single pieces of porous material formed by arrays of interconnected polymer globules or silica skeletons intercalated by interconnected large flow-through pores (1 – 2 μm). These features enable separations at high flow rates and low back pressures. Organic (Svec and Huber 2006), inorganic (Ishizuka et al. 2000, 2002), and hybrid organic-inorganic materials (Kabir, Furton, and Malik 2013; Zhu and Row 2012; Alves, Scholder, and Nischang 2013) are currently the most common monoliths used for LC and SPE (Masini, do Nascimento, and Vitek 2021; Masini and Svec 2017) (Gama, Rocha, and Bottoli 2019).

Organic monoliths can be easily prepared within the confines of a column (mold) by thermal or photo-initiated polymerization of a mixture of monomers (functional and crosslinker), porogenic solvents, and a free radical initiator. Styrene, acrylamide (AAm), alkyl methacrylates, methacrylic acid (MAA), 4-vinylpyridine (4-VP) and glycidyl methacrylate (GMA) are typical functional monomers, whereas ethylene glycol dimethacrylate (EDMA) and divinylbenzene (DVB) are the most often used crosslinkers (Guiochon 2007; Svec and Huber 2006). Organic polymer-based monoliths efficiently retain and separate large molecules such as proteins, peptides, nucleic acids, viruses, and synthetic polymers. Their porous structure reduces flow resistance and enables elution using high flow rates and steep gradients, as demonstrated in several articles and reviews (Guiochon 2007; Hong et al. 2016; Svec and Huber 2006; Svec and Lv 2015).

On the other hand, the organic monoliths are not efficient in retaining and separating low molar mass analytes. One possible explanation is their low surface area and absence of mesopores, which differentiate them from silica and hybrid monoliths. Another explanation is that their pore surface features a gel structure into which slow diffusion controls the mass transport of the analytes towards the interface between the mobile and stationary phases (Nischang 2013; Causon and Nischang 2014). Recent research focuses on improvements in the separation efficiency of polymeric monoliths for small molecules since these monoliths are easier to prepare compared to their silica counterparts and have better chemical stability enabling their use at extreme pH values (Svec 2012; Lin et al. 2015; Urban, Jandera, and Langmaier 2011; Jandera et al. 2013; Urban, Svec, and Fréchet 2010). Svec and Lv (2015) and Hong et al. (2016) reviewed the new approaches concerning the preparation of monolithic columns. These new approaches include, for instance, increasing surface areas using hybrid materials, incorporating nanoparticles and metal-organic frameworks (MOF).

Besides their stability in a wide pH range, these polymers can be prepared in the confines of stainless steel (Svec 2010), fused silica (Svec and Huber 2006; Guiochon 2007; Svec and Lv 2015a), fused silica lined stainless steel (Umemura et al. 2006; Masini 2016b, 2016a), poly(ether-ether-ketone) (PEEK) (Leito, Lv, and Heiter 2016; Shu et al. 2012), and polypropylene (PP) tubes (Iacono, Connolly, and Heise 2016; do Nascimento et al. 2020). PP devices are the cheaper and more versatile among these supports because they are readily available in the conical format of pipette tips, syringe barrels typically used in offline SPE devices, and spin columns, besides straight tubes (Iacono, Connolly, and Heise 2016).

The UV transparency of PP materials makes them compatible with photopolymerization, an attractive approach for the fast construction of polymeric monoliths. While thermal polymerization takes several hours to achieve an acceptable reaction yield, photopolymerization requires only a few minutes. Additionally, PP tubes are mechanically resistant, enabling the use of moderate pressures (Stachowiak et al. 2003; Iacono, Connolly, and Heise 2016; Catalá-Icardo et al. 2018, 2020).

This chapter describes the procedure to obtain stable monolithic beds anchored in PP surfaces and highlights the recent application of these devices in several analytical and bioanalytical chemistry fields.

Polypropylene Functionalization

The adsorbent inside the chromatographic column or SPE cartridge must be physically stable to avoid channeling or void spaces between the sorbent and the support. Voids and channeling contribute to the dead volume and worsens the retention/efficiency of separation. Filters and frits are necessary to retain the particles inside the support. Monoliths, however, can covalently bind to the inner walls of PP tubes producing very stable sorbent beds. The bonding prevents flow through the void between the monolith and wall. Additionally, the binding to the column end offers the advantages of no dead volume and eliminates the use of frits, which are a known source of secondary adsorption mechanisms (Blomberg 2009).

The grafting of pendant vinyl groups to the inner walls of the PP housing enables the covalent binding of internal tube walls to the monolith, thus avoiding channeling and the detachment from the tube under pressure exerted by the mobile phase in the case of straight cylindrical tubes. Vinylization of PP relies on a two-step approach, proposed by Stachowiak et al. in 2003 (Stachowiak et al. 2003). In the first step, benzophenone (BP) is photografted on the PP walls to create a surface photoinitiator. In this step, photochemically produced triplet states of BP carbonyl groups abstracts H atoms from the PP methyl groups, producing fast recombining semi-pinacolic radicals and surface-bound dormant radicals regenerated in the reaction with a reactive monomer such as EDMA in the second step to form a reticulated poly(EDMA) layer plenty of pendant vinyl groups (Figure 1).

A 5 wt% BP in methanol warrants total surface coverage of PP surface upon UV radiation for 20 min under 254 nm at 120 mJ/cm^2 (do Nascimento et al. 2020). Recent studies suggest the tubes should be 2.5 cm apart from the UV lamps (Henrique do Nascimento et al. 2020; Catalá-Icardo et al. 2018) to achieve efficient grafting. Castell et al. reported that BP grafted PP surfaces are stable for three weeks in the dark (Castell et al. 2004).

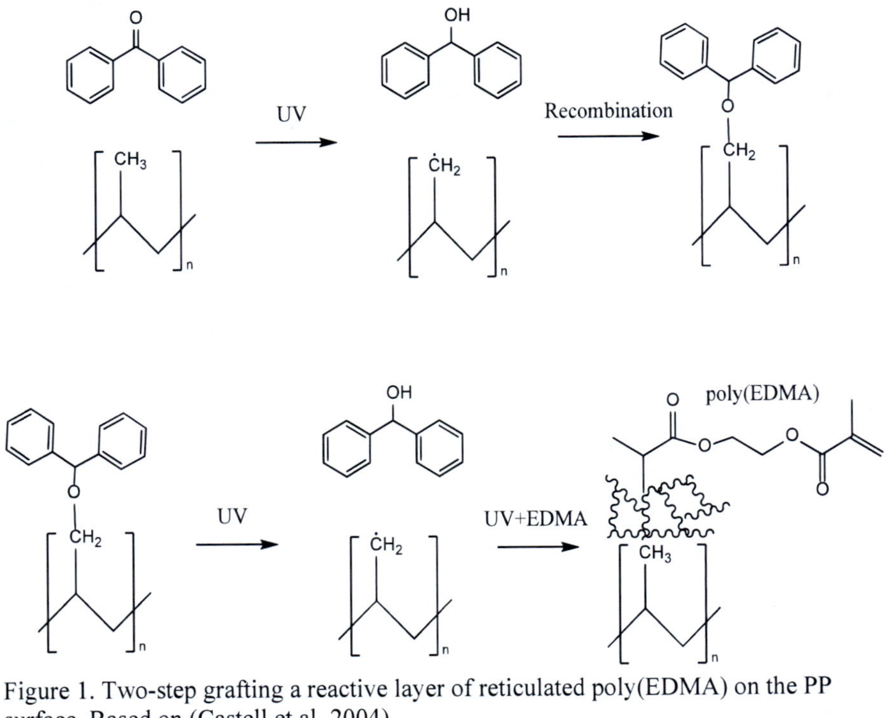

Figure 1. Two-step grafting a reactive layer of reticulated poly(EDMA) on the PP surface. Based on (Castell et al. 2004).

After this first step, the tubes are washed with methanol and dried with N_2, thus removing the unreacted BP. In the second step, the tubes filled with a sonicated and deaerated solution of 15 wt% EDMA in methanol are irradiated for 20 min under 254 nm at 120 mJ/cm^2. The tubes are then washed with methanol and dried under a flow of N_2. In this second step, the BP initiator mediates the photografting of a thin layer of cross-linked poly(EDMA), thus forming an assortment of terminal vinyl groups useful to anchor the monolith inside the housing. The poly(EDMA) layer works as an intermediate ligand between the PP wall and the polymer monolith, formed in the presence of free-radical initiators like azobisisobutyronitrile (AIBN) and 2,2-dimethoxy-2-phenyl acetophenone (DMPAP). The copolymers form simultaneously with their covalent binding to the pendant vinyl groups of the grafted poly(EDMA) layer.

Characterization of the modified PP surface by attenuated total reflectance Fourier transform infrared (ATR-FTIR) spectroscopy shows, after the first modification step, a reflectance peak at 703 cm^{-1} (absent in the bare PP) due to the immobilized semi-pinacolic form of the BP radical (Iacono, Connolly, and Heise 2016). This reflectance peak increases up to 15-20 min of irradiation when it stabilizes due to saturation and starts to decrease, probably because of reaction with atmospheric O_2. Other peaks that confirm first-step modification are the O-H stretching band at 3423 cm^{-1} and an absorption centered at 1628 cm^{-1} due to aromatic C=C, confirming the unsaturation of the surface of PP after the BP grafting.

In the second step of EDMA photografting, 20 min irradiation suffices to cause the disappearance of the bands centered at 3423 cm^{-1} and 1628 cm^{-1}, remaining a small and narrower band at 1630 cm^{-1}, which corresponds to C=C stretching vibrations of the grafted poly(EDMA). However, the primary evidence of poly(EDMA) grafting is the appearance of an intense band at 1730 - 1734 cm^{-1} due to C=O stretching of the carbonyl groups. Scanning electron microscopy is another characterization technique. Figure 2 shows the interfaces between the tube walls and the polymer monoliths without void spaces between them.

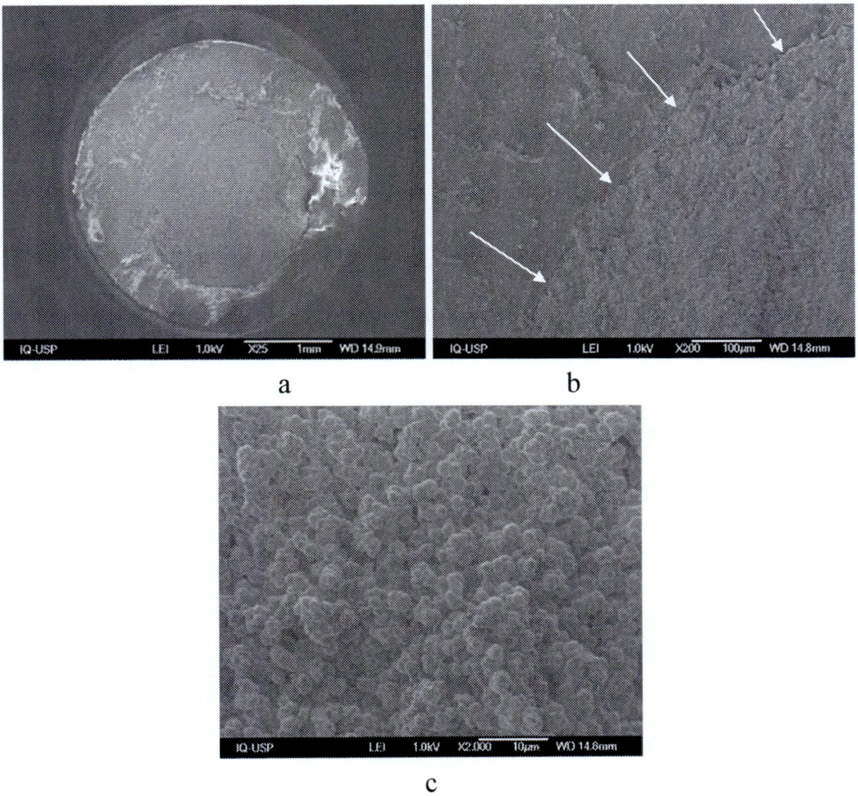

Figure 2. SEM images of (a) the radial end of ink-pen PP straight tube housing a poly(GMA-co-EDMA)-SO_3 monolith; (b) the firm attachment of the polymer to the PP inner walls, with the arrows emphasizing the interface, and (c) the core of the monolithic copolymer with its interconnected globules and flow-through pores.

Polypropylene Devices and Applications

Namera et al. (2011) reviewed the different online and offline monolithic SPE applications in drugs and pharmaceutical analysis, describing several devices constructed in PP such as pipette tips, syringe filter, and spin columns.

Table 1. Examples of PP devices, polymer monoliths for applications in Solid Phase Extraction (SPE), Solid Phase Microextraction (SPME), Molecularly Imprinted Solid Phase Extraction (MISPE)

PP Device	Polymer	Functionalization	Extraction Technique	Sample	Analytes	Analytical Technique	Recovery (%)	LOD	Reference
Pipette tip	Poly(MMA-co-EDMA)	-	SPME	Swine and cattle urine	ractopamine and clenbuterol	HPLC-UV and UHPLC Orbitrap HRMS	92 (ractopamine) 100 (clenbuterol)	3.75 (ractopamine) and 1.88 (clenbuterol) ng/mL - HPLC-UV 0.34 (ractopamine) and 0.19 (clenbuterol) ng/mL - UHPLC	(Ar-Sanork et al. 2021)
Pipette tip	Poly(4-VP-co-EDMA)	2,4-dichloro-phenoxyacetic acid as template	MISPE	Lake water	2,4-D	HPLC-UV	114	0.002 mg/L	(Wu et al. 2020)
Pipette tip	Poly(MAA-co-EDMA)	-	SPME	Human oral fluids	cocaine, diazepam, methamphetamine, and 20 NPS	UHPLC-MS-MS	64 to 115	0.03 to 0.6 µg/L	(Sorribes-Soriano et al. 2019)
Pipette tip	Poly(GMA-co-EDMA)	Multi-walled carbon nanotubes	SPME	Human urine	antidepressants	HPLC-UV	86±12 (mianserine) 108±16 (desipramine)	9 (desipramine), 13 (mianserine), 15 (amitriptyline),	Fresco-Cala (2018)

PP Device	Polymer	Functionalization	Extraction Technique	Sample	Analytes	Analytical Technique	Recovery (%)	LOD	Reference
							83±15 (amitriptyline) 93±15 (trimipramine)	15 (trimipramine) µg/L	
Pipette tip	Acryloyl β-cyclodextrin hybrid-silica hybrid	-	SPME	Lettuce	methyl parathion and fenthion	HPLC-UV	96.0 to 104.2	4.5 (methyl parathion) and 6.0 µg/kg (fenthion)	(Chen et al. 2018)
Pipette tip	Poly(MMA-co-EDMA)	Tetrapeptide (Tyr-Gly-Gly-Phe) as the template	MISPE	Human cerebrospinal fluid	enkephalins	HPLC-UV	86.1-93.0 (met-enkephalin) and 90.4-95.7 (enkephalin)	0.08 (met-enkephalin) and 0.05 (leu-enkephalin) nmol/L	(Li and Li 2015)
Pipette tip	Poly(MMA-co-EGDMA)	Methomyl template	MISPE	Environmental and wastewater	methomyl	HPLC-UV	84.9 to 103.6	0.2 µg/L	(Du et al. 2014b)
Pipette tip	Poly(MAA-co-EDMA)	Difenoconazole template	MISPE	Water and grape juice	difenoconazole	HPLC-UV	89.3 – 95.4 (water) 87.6 – 97.9 (juice)	0.5 µg/L	(Du et al. 2014a)
Syringe barrel	Poly(GMA-co-EDMA)	Cystamine-TCEP-AuNP-PMPMS-VBPA	SPE	Human serum and HRP digests	glycopeptides	Nano-LC-MS-MS	-	0.5 fmol/µL	(Mompó-Roselló et al. 2021)
Syringe barrel	Poly(GMA-co-EDMA)	Amino and thiol-groups followed by AuNPs	SPME	Human saliva and urine	glutathione	HPLC-fluorometric detection	86 to 105	1.5 ng/mL	(Mompó-Roselló, Ribera-Castelló, et al. 2020)

Table 1. (Continued)

PP Device	Polymer	Functionalization	Extraction Technique	Sample	Analytes	Analytical Technique	Recovery (%)	LOD	Reference
Syringe barrel	Poly(GMA-co-EDMA)	Ethylenediamine	SPE	Wastewater	Pb^{2+}, Cu^{2+}, Cr^{3+} and Ni^{2+}	AAS	98	-	
Spin column	Poly(GMA-co-EDMA)	Imidazolium-based ionic liquid	SPME	Human urine	acebutolol, carteolol, oxprenolol, alprenolol and propranolol	HPLC-UV	>90	1.4 to 40 μg/L	(Mompó-Roselló, Ribera-Castelló, et al. 2020)
Spin column	Poly(GMA-co-EDMA)	MOF NH_2-MIL-101(Cr)	SPE	Human urine	Ketoprofen, ibuprofen and flurbiprofen	HPLC-UV	71.0 to 78.0	0.1 to 7 μg/L	(Giesbers et al. 2019)
Syringe Filter tips	Poly(GMA-co-EDMA)	Iminodiacetic acid	SPME	River waters	Ti, Fe, Co, Ni, Cu, Ga, Cd, Sn and REEs	ICP-MS	>80	$3x10^{-6}$ to 0.18 g/L	(Rahmi et al. 2010)

This chapter presents (Table 1) some of the most recent applications of PP devices accommodating polymeric monoliths, describing their main characteristics and the respective analytical features such as recoveries and limits of detection (LOD).

Ink-Pen Tubes

Iacono, Connolly, and Heise (2016) described the functionalization of PP tubes obtained from ink pens to construct 2.0 mm id poly(glycidyl methacrylate-co-ethylene glycol dimethacrylate), poly(GMA-co-EDMA), columns. They demonstrated the efficient monolith attachment to the tube wall and the structural stability to flow rates of 500 μL/min of methanol for at least 8-h. This work was proof of concept to demonstrate the potential usability of straight PP tubes to house monolithic columns but did not demonstrate any analytical application.

Do Nascimento et al. (2020) prepared poly(GMA-co-EDMA) columns following the procedures of Iacono et al. (2016) and then functionalized the parent monolith to prepare cation and anion exchange columns (Ribeiro et al. 2018). For this preparation, a methanolic solution of benzophenone (5 wt%) activated the inner walls of the ink-pen tubes (measuring 100-110 mm long and 2 mm internal diameter) to covalently bind with crosslinker EDMA, being all procedures assisted by UV. After the inner wall functionalization, photopolymerization of poly(GMA-co-EDMA) proceeded simultaneously with the binding of the copolymer to the pendant vinyl groups. Flushing the column with acetonitrile removed the unreacted monomers, porogenic solvents, and initiator. The last washing step with deionized water removed the acetonitrile. Functionalization of the epoxy groups in the parent poly(GMA-co-EDMA) monoliths with iminodiacetate (IDA) and Na_2SO_3 produced weak and strong cation exchangers, respectively. These monoliths resisted flow rates up to 2.0 mL/min of water (back pressures between 6.9 and 9.0 MPa) without any evidence of structure collapse, tube wall rupture, or monolith detachment.

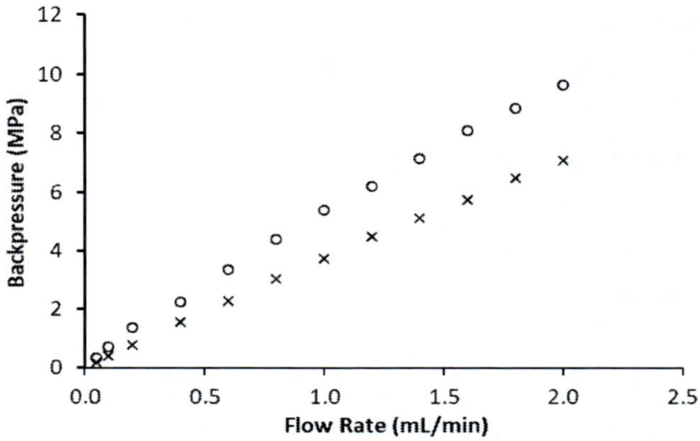

Figure 3. Effect of the flow rate on the backpressure of (o) poly(GMA-co-EDMA)-SO_3 and (×) poly(GMA-co-EDMA)-IDA in 100 mm long × 2 mm i.d. column. The mobile phase was deionized water at ambient temperature. At flow rates > 2 mL/min, leakage appeared in the column connections.

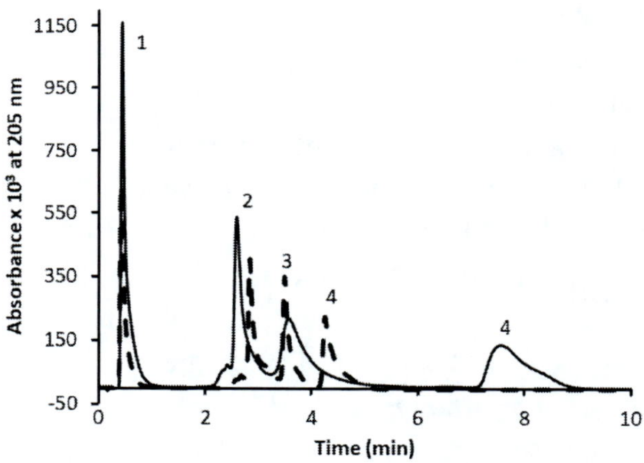

Figure 4. Blank subtracted cation exchange chromatograms of (1) ovalbumin, (2) ribonuclease A, (3) cytochrome C, and (4) lysozyme on poly(GMA-*co*-EDMA)-SO_3 (dashed line) and poly(GMA-co-EDMA)-IDA (continuous line). All proteins at a concentration of 1.0 mg/mL. Sample volume = 2 μL, flow rate = 500 μL/min, temperature = 60°C. Elution program in the poly(GMA-co-EDMA)-SO_3: isocratic at 0% **B** from 0 to 2 min; linear gradient from 0 to 30% **B** in 2 min; isocratic at 30% **B** from 4 to 6 min; linear gradient from 30 to 0% **B** in 0.1 min; isocratic at 0% B until 10 min. Elution program in the poly(GMA-co-EDMA)-IDA: isocratic at 0% **B** from 0 to 0.5 min; linear gradient from 0 to 10% **B** in 2.5 min; linear gradient from 10 to 20% **B** in 4 min; linear gradient from 20 to 0% **B** in 0.1 min; isocratic at 0% **B** from 7.1 to 11 min. Mobile Phase **A**: 20 mM phosphate buffer (pH 7.0); Mobile Phase **B**: 1.0 M NaCl in 20 mM phosphate buffer (pH 7.0)

They exhibited permeabilities between 2.7 to 3.3×10^{-13} m^2, enabling the separation of ovalbumin, ribonuclease A, cytochrome C, and lysozyme in salt gradient at pH 7.0 following the order of increasing isoelectric points, thus confirming the cation exchange mechanism. During these separations, the working flow rate was 500 µL/min (resulting in backpressures less than 2.8 MPa). Determination of lysozyme in egg white demonstrated the applicability of the columns in the analysis of complex samples.

Easy, cheap, and fast preparation of polymer monoliths within the confines of ink-pen PP tubes are the most attractive features of this approach. Pressure-resistance tests associated with characterizations by SEM and ATR-FTIR proved the firm adhesion of the monolith to the inner walls of the straight tubes. In addition, the modification of the parent monolith with various reagents without affecting the monolith stability and its covalent binding to the PP surface confirms the potential applicability of these columns in moderate pressure liquid chromatography and automated solid-phase extraction.

Syringe Barrels

SPE cartridges are typically used for sample preparation before chromatographic analysis, aiming to clean up the sample by eliminating coexisting substances, minimize matrix interferences, and concentrate the analytes. The most common SPE technique uses syringe cartridges packed with 100 to 1000 mg of the particulate (~ 40 µm) stationary phase affording high affinity for the analytes. Two PP frits with 20 µm pores, one at the bottom of the syringe and the other at the top of the stationary phase, keep the sorbent bed stabilized. Compared with conventional liquid-liquid extraction, SPE in cartridges reduces sample handling and offers significant time-saving in the sample preparation besides reducing solvents consumption and analyst exposure. SPE cartridges with diverse chemistries are available on the market, affording ion-exchange, reversed-phase, hydrophilic interaction retention mechanisms, besides more sophisticated materials with molecular recognition mechanisms as those used in immunosorbents and molecularly imprinted polymers (MIP). Their costs are relatively high. Thus, alternatively, stationary phases for SPE can be prepared in the laboratory exploring the polymeric monoliths based on acrylates and using syringe barrels as the mold. The use of polymer monoliths circumvents some limitations of conventional cartridges for SPE, such as the use in conditions of $2.5 < pH < 8$ (especially in the case of silica-based sorbents), or materials that have slow diffusional mass transport, or materials that generate high backpressure at high flow rates (Aguiar et al. 2018; H. Wang et al. 2014; Pandey and Kim 2012). Besides, the covalent binding of the monolith to the PP internal walls dispenses the use of frits to accommodate and stabilize the sorbent bed.

Wang et al. (2014) prepared the parent poly(GMA-co-EDMA) monolith in a PP syringe barrel and functionalized the epoxy groups of GMA with the chelating ethylenediamine to concentrate toxic metals (Pb^{2+}, Ni^{2+}, Cu^{2+}, Cr^{3+}) with up to 300-fold enrichment factor. Mompó-Roselló, Vergara-Barberán, et al. (2020) prepared 20 mg of poly(GMA-co-EDMA) monolith by photopolymerization inside a 1-mL vinylized PP syringe. They then functionalized the parent monolith with ammonia, cysteamine, or cystamine to form a pore surface rich in amine or thiol groups able to attach gold nanoparticles (AuNP). Since the surface of the AuNP remains active after immobilization and has a high affinity for thiolated compounds (Ribeiro, Masini, and Svec 2019; Xu et al. 2010), the SPE device extracted glutathione, an antioxidant whose depletion in biological fluids suggests aging and diseases such as arthritis, muscular dystrophy, and Alzheimer's disease. The SPE of glutathione from saliva achieved an effective sample clean-up and enrichment factors between 86 and 105% (Table 1).

Mompó-Roselló et al. demonstrated the proteomics applicability of SPE cartridges constructed in PP syringes (Mompó-Roselló et al. 2021). They prepared poly(GMA-co-EDMA) inside the syringe barrels and then functionalized the parent polymer by two distinct strategies. In the first one, poly-3-mercaptopropyl-methylsiloxane (PMPMS) reacted with the surface epoxy groups leaving several free thiols to react with vinyl phenylboronic acid (VPBA) via thiol-ene reaction to form a boronate affinity sorbent. In the second approach, the strategy first described by Lv et al. (2012) based on the functionalization of the epoxy groups with cystamine and then the cleavage of the disulfide bond by tris(2-carboxyethyl)phosphine hydrochloride (TCEP) produced a high density of free thiols to immobilize AuNPs which served as intermediate ligands to attach PMPMS and then VPBA. The second approach provided more abundant boronic acid groups at the pore surface for efficient enrichment of glycopeptides in a mixture of tryptic digests of horse-radish peroxidases (HRP) and bovine serum albumin (BSA).

As demonstrated in these examples, the SPE cartridges constructed in syringe barrels and polymeric monoliths have found an outstanding variety of applications. Simplicity, cost-effectiveness, chemical versatility, and reusability are advantageous features of these devices compared to the conventional SPE cartridges. Furthermore, adequate modification of polymeric monoliths contemplates various applications (biological, environmental, pharmaceutical).

Pipette Tips

The conical format of pipette tips favors the packing of the sorbent material, exploring the funneling walls of the tip to hold stable the extractant phase without any wall functionalization. However, such a configuration works better in a unidirectional flow of conditioner, sample, washer, and eluant solutions with the aid of syringe pumps (Wang et al. 2020). The resources of automatic or manual pipettes to perform both aspiration and injection cycles should be used with care because the aspiration may displace the sorbent bed or the monolith upward, creating void volumes between the bed and the tube walls (Chi et al. 2019). The covalent binding of the monolith to the PP walls avoids detachment and channeling during the several aspiration/pushing cycles for the conditioning, sampling, washing, and eluting steps needed in SPE procedures. Stachowiak et al. pioneered the use PP pipette tips grafted with a thin layer of ethylene diacrylate to house monolithic structures (Stachowiak et al. 2003). The success of this approach led to the use of monolith plugs in PP tips in 96-well plate systems, primarily used in sample preparation for bioanalysis, as reviewed by Blomberg back in 2009 (Blomberg 2009).

In more recent years, Sorribes-Soriano et al. described the construction of poly(methacrylic acid-co-ethylene glycol dimethacrylate), poly(MAA-co-EDMA), inside 200 μL pipette tips for the extraction of drugs of abuse from oral fluid samples (Sorribes-Soriano et al. 2019). The in-tip monolith extracted cocaine, diazepam, methamphetamine, and twenty new psychoactive substances (NPS) extracted from oral fluids using an eight-channel micropipette before the analysis by ultra-high-performance liquid chromatography-tandem mass spectrometry. Table 1 shows other applications in biological samples.

Molecularly imprinted solid-phase extraction (MISPE) increases the selectivity of conventional SPE. This technique relies on synthesizing the polymer monolith in the presence of the target template to create molecular recognition cavities in the sorbent. Wu et al. (2020) manufactured a tip extraction device by co-sintering at 150°C, inside a PP pipette tip, high-density polyethylene powder and a MIP formed between the functional monomer 4-vinyl pyridine (4-VP), the crosslinker EDMA and the template 2,4-dichlorophenoxyacetic acid (2,4 D) by free-radical copolymerization in a methanol/water (4/1, v/v) using AIBN as initiator. The formed monolith removed 2,4-D from lake water samples with an enrichment factor of around 50 (Table 1).

The poly(MAA-co-EDMA) generic monolith copolymerized inside pipette tips in the presence of methomyl (a pesticide) or difenoconazole (fungicide) served for MISPE of difenoconazole (Du et al. 2014a) from waters and grape juice and methomyl from environmental waters (Du et al. 2014b) after the template removal (Table 1). In these cases, the authors did not functionalize the tip walls so that unidirectional flow driven by syringe pumps performed all the conditioning, extraction, washing, and elution steps.

Many examples show the solid-phase microextraction in pipette tips attends the demand for low-cost, miniaturized schemes of sample preparation, using small volumes (μL range) of conditioning, sample, and eluent solutions, enabling rapid and quantitative extractions with possible reuse of devices, as demonstrated by (Chen et al. 2018; Ar-Sanork et al. 2021; Wu et al. 2020; Fresco-Cala et al. 2018).

Spin Columns

SPE in spin columns offers advantages of enabling simultaneous processing of multiple samples, limited by the number of positions of the centrifuge, and low consumption of solvent (Giesbers et al. 2019; Mompó-Roselló, Vergara-Barberán, et al. 2020). An advantage of the spin column extraction technique is that all extraction devices suffer the same centrifuge force, leading to an excellent reproducibility of the flow rate in both the loading and eluting steps. The uniformity in flow rate in multiposition manifolds operating under vacuum with conventional SPE cartridges is not as good as that of spin columns. High repeatability of flow rate results in improved repeatability of recoveries and analytical results.

Combining centrifugation and extracting polymeric monoliths with the appropriate chemical functionalization created a simple, fast, and attractive alternative for sample treatment, with great application potential, especially in bioanalysis. Mompó-Roselló et al. prepared a small volume (70 μL) poly(GMA-co-EDMA) monolith in vinylized PP centrifugal devices. They functionalized the pore surfaces with the ionic liquid 1-allyl-3-methylimidazolium chloride to extract beta-blockers from water and urine samples. The ionic liquid enhanced the retention of the analytes by providing electrostatic mechanisms, besides hydrophobic, π-π stacking, and hydrogen bonding interaction. All SPE steps were done by centrifugation, thus significantly reducing costs and time in sample treatment. The devices were quickly prepared and enabled handling many samples simultaneously (>40 samples/hour). Under optimal conditions, β-blockers were quantitatively retained in the modified monolith at pH 12 and desorbed with a water-methanol mixture (Mompó-Roselló, Ribera-Castelló, et al. 2020).

Aiming to enrich nonsteroidal anti-inflammatory drugs (NSAIDs) from urine and water samples, Giesbers et al. prepared poly(GMA-co-EDMA) monoliths inside vinylized PP spin columns. They functionalized the polymer with the metal-organic framework (MOF) NH_2-MIL-101 (Cr) by two strategies: (i) the NH_2-MIL-101 (Cr) was dispersed in the polymerization mixture (0.1 to 1 wt%), embedding the MOF in the polymer, and (ii) the MOF's amino groups reacted with the preformed poly(GMA-co-EDMA) epoxy moieties, thus functionalizing only the pore surfaces. The second approach enhanced the availability of sorption sites and provided retention of ketoprofen, flurbiprofen, and ibuprofen approaching 100%, while both the parent and the MOF-embedded monolith removed less than 40% of the target compounds. Multiple interaction mechanisms such as hydrogen bonding, π-π stacking, and hydrophobic interactions provided by MOF particles explained the high recoveries and enabled the quantification of the NSAIDs in waters and urine (Table 1) (Giesbers et al. 2019).

Conclusion

This chapter presented some uses of PP laboratory devices to accommodate polymer monoliths for uses in SPE and chromatographic separations. Cheap PP devices with high mechanical and chemical stability enabled their use to accommodate extractant materials for sample preparation in analytical methods exploring a wide range of chemistries (ion-exchange, hydrophobic interactions, molecular recognition of MIPs, and multiple mechanisms in MOFs). Despite the PP chemical stability, it can be easily and quickly functionalized by photopolymerization of a thin layer of poly(EDMA) to provide a high density of pendant vinyl groups that enable the covalent binding of the stationary phase to the tube walls. This binding keeps the stationary phase stable in the support and is essential for constructing very cheap chromatographic columns to separate proteins by reversed-phase and ion-exchange liquid chromatography. Ongoing research aims to develop polymer and hybrid monolithic columns inside PP tubes to separate and concentrate small molecules.

ACKNOWLEDGMENTS

Research on monolithic materials has been funded by grants 2013/18507–4 from the São Paulo Research Foundation (FAPESP) and 303940/2017–4 from the National Council for Scientific and Technological Development (CNPq). FHN acknowledges Coordination for the Improvement of Higher Education Personnel (CAPES) for a post-doc fellowship (Contract 88882.315696/2019-01).

REFERENCES

Aguiar, Carlos A. S., Nadja T. M. Silva, Dayane F. Soares, and Anizio M. Faria. 2018. "Solid-Phase Extraction of Pesticide Residues from Surface Water Using a Lab-Made Lower-Retention Sorbent." *Quimica Nova* 41 (6): 641–47. https://doi.org/10.21577/0100-4042.20170223.

Alves, Filipa, Pascal Scholder, and Ivo Nischang. 2013. "Conceptual Design of Large Surface Area Porous Polymeric Hybrid Media Based on Polyhedral Oligomeric Silsesquioxane Precursors: Preparation, Tailoring of Porous Properties, and Internal Surface Functionalization." *ACS Applied Materials and Interfaces* 5 (7): 2517–26. https://doi.org/10.1021/am303048y.

Ar-Sanork, Kesara, Chanpen Karuwan, Nakin Surapanich, Prapin Wilairat, Duangjai Nacapricha, and Patcharin Chaisuwan. 2021. "Mixed Mode Monolithic Sorbent in Pipette Tip for Extraction of

Ractopamine and Clenbuterol Prior to Analysis by HPLC-UV and UHPLC-Q ExactiveTM Plus Orbitrap MS." *Journal of Analytical Science and Technology* 12:23 (1): 1–10. https://doi.org/10.1186/s40543-021-00275-5.

Blomberg, Lars G. 2009. "Two New Techniques for Sample Preparation in Bioanalysis: Microextraction in Packed Sorbent (MEPS) and Use of a Bonded Monolith as Sorbent for Sample Preparation in Polypropylene Tips for 96-Well Plates." *Analytical and Bioanalytical Chemistry* 393 (3): 797–807. https://doi.org/10.1007/s00216-008-2305-4.

Castell, P., M. Wouters, G. De With, H. Fischer, and F. Huijs. 2004. "Surface Modification of Poly(Propylene) by Photoinitiators: Improvement of Adhesion and Wettability." *Journal of Applied Polymer Science* 92 (4): 2341–50. https://doi.org/10.1002/app.20276.

Catalá-Icardo, M., S. Torres-Cartas, S. Meseguer-Lloret, E. F. Simó-Alfonso, and J. M. Herrero-Martínez. 2018. "Photografted Fluoropolymers as Novel Chromatographic Supports for Polymeric Monolithic Stationary Phases." *Talanta* 187 (March): 216–22. https://doi.org/10.1016/j.talanta.2018.05.026.

Catalá-Icardo, M., S. Torres-Cartas, E. F. Simó-Alfonso, and J. M. Herrero-Martínez. 2020. "Influence of Photo-Initiators in the Preparation of Methacrylate Monoliths into Poly(Ethylene-Co-Tetrafluoroethylene) Tubing for Microbore HPLC." *Analytica Chimica Acta* 1093: 160–67. https://doi.org/10.1016/j.aca.2019.09.055.

Causon, Tim J, and Ivo Nischang. 2014. "Critical Differences in Chromatographic Properties of Silica- and Polymer-Based Monoliths." *Journal of Chromatography A* 1358: 165–71. https://doi.org/10.1016/j.chroma.2014.06.102.

Chen, Ling, Xueping Dang, Youhong Ai, and Huaixia Chen. 2018. "Preparation of an Acryloyl β-Cyclodextrin-Silica Hybrid Monolithic Column and Its Application in Pipette Tip Solid-Phase Extraction and HPLC Analysis of Methyl Parathion and Fenthion." *Journal of*

Separation Science 41 (18): 3508–14. https://doi.org/10.1002/jssc.201701273.

Chi, Shuai Shuai, Chao Peng, Yao Han Lan, Hui Yuan Yang, Xin Zhu Chen, Luo Bing Han, Chao Zheng, Lin Yi Dong, and Xian Hua Wang. 2019. "Preparation of Phenyl-Boronic Acid Polymer Monolith by Initiator-Free Ring-Opening Polymerization for Microextraction of Sulfamethoxazole and Trimethoprim from Animal-Originated Foodstuffs." *Journal of Chromatography A* 1590: 10–18. https://doi.org/10.1016/j.chroma.2018.12.067.

Du, Ting, Jing Cheng, Min Wu, Xiaohua Wang, Hongbin Zhou, and Min Cheng. 2014a. "An in Situ Immobilized Pipette Tip Solid Phase Microextraction Method Based on Molecularly Imprinted Polymer Monolith for the Selective Determination of Difenoconazole in Tap Water and Grape Juice." *Journal of Chromatography B: Analytical Technologies in the Biomedical and Life Sciences* 951–952 (1): 104–9. https://doi.org/10.1016/j.jchromb.2014.01.030.

Du, Ting, Jing Cheng, Min Wu, Xiaohua Wang, Hongbin Zhou, Min Cheng. 2014b. "Pipette Tip-Based Molecularly Imprinted Monolith for Selective Micro-Solid-Phase Extraction of Methomyl in Environmental Water." *Analytical Methods* 6 (16): 6375–80. https://doi.org/10.1039/c4ay00766b.

Fresco-Cala, Beatriz, Óscar Mompó-Roselló, Ernesto F. Simó-Alfonso, Soledad Cárdenas, and José Manuel Herrero-Martínez. 2018. "Carbon Nanotube-Modified Monolithic Polymethacrylate Pipette Tips for (Micro)Solid-Phase Extraction of Antidepressants from Urine Samples." *Microchimica Acta* 185 (2): 1–7. https://doi.org/10.1007/s00604-017-2659-4.

Gama, Mariana R., Fábio R. P. Rocha, and Carla B. G. Bottoli. 2019. "Monoliths: Synthetic Routes, Functionalization and Innovative Analytical Applications." *TrAC - Trends in Analytical Chemistry* 115: 39-51. https://doi.org/10.1016/j.trac.2019.03.020.

Giesbers, Myrthe, Enrique Javier Carrasco-Correa, Ernesto F. Simó-Alfonso, and José Manuel Herrero-Martínez. 2019. "Hybrid Monoliths with Metal-Organic Frameworks in Spin Columns for

Extraction of Non-Steroidal Drugs Prior to Their Quantitation by Reversed-Phase HPLC." *Microchimica Acta* 186 (12): 1–9. https://doi.org/10.1007/s00604-019-3923-6.

Guiochon, Georges. 2007. "Monolithic Columns in High-Performance Liquid Chromatography." *Journal of Chromatography A* 1168 (1–2): 101–68. https://doi.org/10.1016/j.chroma.2007.05.090.

do Nascimento, Fernando, Catharina Raquel Lemos Trazzi, Amanda Hanashiro Moraes, Caryna Moraes Velasques, Diego Miranda de Souza Costa, and Jorge Cesar Masini. 2020. "Construction of Polymer Monolithic Columns in Polypropylene Ink-Pen Tubes for Separation of Proteins by Cation-Exchange Chromatography." *Journal of Separation Science* 43 (22): 4123–30. https://doi.org/10.1002/jssc.202000803.

Hong, Tingting, Xi Yang, Yujing Xu, and Yibing Ji. 2016. "Recent Advances in the Preparation and Application of Monolithic Capillary Columns in Separation Science." *Analytica Chimica Acta* 931: 1–24. https://doi.org/10.1016/j.aca.2016.05.013.

Iacono, Marcello, Damian Connolly, and Andreas Heise. 2016. "Fabrication of a GMA-Co-EDMA Monolith in a 2.0 Mm i.d. Polypropylene Housing." *Materials* 9 (4): 263. https://doi.org/ 10.3390/ma9040263.

Ishizuka, Norio, Hiroshi Kobayashi, Hiroyoshi Minakuchi, Kazuki Nakanishi, Kazuyuki Hirao, Ken Hosoya, Tohru Ikegami, and Nobuo Tanaka. 2002. "Monolithic Silica Columns for High-Efficiency Separations by High-Performance Liquid Chromatography." *Journal of Chromatography A* 960 (1–2): 85–96. https://doi.org/10.1016/ S0021-9673(01)01580-1.

Ishizuka, Norio, Hiroyoshi Minakuchi, Kazuki Nakanishi, Naohiro Soga, Hisashi Nagayama, Ken Hosoya, and Nobuo Tanaka. 2000. "Performance of a Monolithic Silica Column in a Capillary under Pressure- Driven and Electrodriven Conditions." *Analytical Chemistry* 72 (6): 1275–80. https://doi.org/10.1021/ac990942q.

Jandera, Pavel, Magda Staňková, Veronika Škeříková, and Jiří Urban. 2013. "Cross-Linker Effects on the Separation Efficiency on

(Poly)Methacrylate Capillary Monolithic Columns. Part I. Reversed-Phase Liquid Chromatography." *Journal of Chromatography A* 1274: 97–106. https://doi.org/10.1016/j.chroma.2012.12.003.

Kabir, Abuzar, Kenneth G. Furton, and Abdul Malik. 2013. "Innovations in Sol-Gel Microextraction Phases for Solvent-Free Sample Preparation in Analytical Chemistry." *TrAC - Trends in Analytical Chemistry* 45: 197–218. https://doi.org/10.1016/j.trac.2012.11.014.

Leito, Ivo, Chunguang Lv, and Jaana Heiter. 2016. "Analytica Chimica Acta Covalent Attachment of Polymeric Monolith to Polyether Ether Ketone (PEEK) Tubing" 932: 114–23. https://doi.org/10.1016/j.aca.2016.05.026.

Li, Hua, and Dan Li. 2015. "Preparation of a Pipette Tip-Based Molecularly Imprinted Solid-Phase Microextraction Monolith by Epitope Approach and Its Application for Determination of Enkephalins in Human Cerebrospinal Fluid." *Journal of Pharmaceutical and Biomedical Analysis* 115: 330–38. https://doi.org/10.1016/j.jpba.2015.07.033.

Lin, Shu Ling, Yu Ru Wu, Tzuen Yeuan Lin, and Ming Ren Fuh. 2015. "Preparation and Evaluation of Poly(Alkyl Methacrylate-Co-Methacrylic Acid-Co-Ethylene Dimethacrylate) Monolithic Columns for Separating Polar Small Molecules by Capillary Liquid Chromatography." *Analytica Chimica Acta* 871: 57–65. https://doi.org/10.1016/j.aca.2015.02.015.

Lv, Yongqin, Fernando Maya Alejandro, Jean M J Fréchet, and Frantisek Svec. 2012. "Preparation of Porous Polymer Monoliths Featuring Enhanced Surface Coverage with Gold Nanoparticles." *Journal of Chromatography A* 1261: 121–28. https://doi.org/10.1016/j.chroma.2012.04.007.

Masini, J. C., F. H. do Nascimento, and R. Vitek. 2021. "Porous Monolithic Materials for Extraction and Preconcentration of Pollutants from Environmental Waters." *Trends in Environmental Analytical Chemistry* 29: e00112. https://doi.org/10.1016/j.teac.2020.e00112.

Masini, Jorge C., and Frantisek Svec. 2017. "Porous Monoliths for On-Line Sample Preparation: A Review." *Analytica Chimica Acta* 964: 24–44. https://doi.org/10.1016/j.aca.2017.02.002.

Masini, Jorge C. 2016a. "Semi-Micro Reversed-Phase Liquid Chromatography for the Separation of Alkyl Benzenes and Proteins Exploiting Methacrylate- and Polystyrene-Based Monolithic Columns." *Journal of Separation Science* 39 (9): 1648–55. https://doi.org/10.1002/jssc.201600049.

Masini, Jorge C. 2016b. "Separation of Proteins by Cation-Exchange Sequential Injection Chromatography Using a Polymeric Monolithic Column." *Analytical and Bioanalytical Chemistry* 408 (5): 1445–52. https://doi.org/10.1007/s00216-015-9242-9.

Mompó-Roselló, Oscar, Ana Ribera-Castelló, Ernesto F. Simó-Alfonso, María José Ruiz-Angel, María Celia García-Alvarez-Coque, and José Manuel Herrero-Martínez. 2020. "Extraction of β-Blockers from Urine with a Polymeric Monolith Modified with 1-Allyl-3-Methylimidazolium Chloride in Spin Column Format." *Talanta* 214 (January): 120860. https://doi.org/10.1016/j.talanta.2020.120860.

Mompó-Roselló, Óscar, María Vergara-Barberán, María Jesús Lerma-García, Ernesto F. Simó-Alfonso, and José Manuel Herrero-Martínez. 2021. "Boronate Affinity Sorbents Based on Thiol-Functionalized Polysiloxane-Polymethacrylate Composite Materials in Syringe Format for Selective Extraction of Glycopeptides." *Microchemical Journal* 164 (January). https://doi.org/10.1016/j.microc.2021.106018.

Mompó-Roselló, Oscar, María Vergara-Barberán, Ernesto F. Simó-Alfonso, and José Manuel Herrero-Martínez. 2020. "In Syringe Hybrid Monoliths Modified with Gold Nanoparticles for Selective Extraction of Glutathione in Biological Fluids Prior to Its Determination by HPLC." *Talanta* 209 (November 2019): 120566. https://doi.org/10.1016/j.talanta.2019.120566.

Namera, Akira, Akihiro Nakamoto, Takeshi Saito, and Shota Miyazaki. 2011. "Monolith as a New Sample Preparation Material: Recent

Devices and Applications." *Journal of Separation Science* 34 (8): 901–24. https://doi.org/10.1002/jssc.201000795.

Nischang, Ivo. 2013. "Porous Polymer Monoliths : Morphology, Porous Properties, Polymer Nanoscale Gel Structure and Their Impact on Chromatographic Performance." *Journal of Chromatography A* 1287: 39–58. https://doi.org/10.1016/j.chroma.2012.11.016.

Pandey, Sudhir Kumar, and Ki Hyun Kim. 2012. "Evaluation of the Solid-Phase Extraction (SPE) Cartridge Method in Combination with Thermal Desorption-Gas Chromatography-Mass Spectrometry (TD-GC-MS) for the Analysis of Different VOCs in Liquid Matrices in Varying PH Conditions." *Journal of Separation Science* 35 (15): 1914–21. https://doi.org/10.1002/jssc.201200044.

Rahmi, Dwinna, Yuka Takasaki, Yanbei Zhu, Hiroharu Kobayashi, Shigeji Konagaya, Hiroki Haraguchi, and Tomonari Umemura. 2010. "Preparation of Monolithic Chelating Adsorbent inside a Syringe Filter Tip for Solid Phase Microextraction of Trace Elements in Natural Water Prior to Their Determination by ICP-MS." *Talanta* 81 (4–5): 1438–45. https://doi.org/10.1016/j.talanta.2010.02.048.

Ribeiro, Luiz F., Renan Lopes Martins, Diego M. de Souza Costa, and Jorge C. Masini. 2018. "Poly Glycidyl Methacrylate-Co-Ethylene Dimethacrylate Porous Monolith as a Versatile Platform for the Development of Separations and Solid-Phase Extractions in Sequential Injection Analyzers." *Journal of Separation Science* 41 (24): 4449–57. https://doi.org/10.1002/jssc.201800995.

Ribeiro, Luiz F., Jorge C. Masini, and Frantisek Svec. 2019. "Use of Thiol Functionalities for the Preparation of Porous Monolithic Structures and Modulation of Their Surface Chemistry: A Review." *TrAC - Trends in Analytical Chemistry* 118: 606–24. https://doi.org/10.1016/j.trac.2019.06.027.

Shu, Shin, Hiroharu Kobayashi, Masaki Okubo, Akhmad Sabarudin, Michio Butsugan, and Tomonari Umemura. 2012. "Chemical Anchoring of Lauryl Methacrylate-Based Reversed Phase Monolith to 1/16' o.d. Polyetheretherketone Tubing." *Journal of*

Chromatography A 1242: 59–66. https://doi.org/10.1016/j.chroma.2012.04.030.

Sorribes-Soriano, A., A. Valencia, F.A. Esteve-Turillas, and J.M. Herrero-Martinez. 2019. "Development of Pipette Tip-Based Poly (Methacrylic Acid-Co-Ethylene Glycol Dimethacrylate) Monolith for the Extraction of Drugs of Abuse from Oral Fluid Samples." *Talanta* 205 (May): 120158. https://doi.org/10.1016/j.talanta.2019.120158.

Stachowiak, Timothy B, Thomas Rohr, Emily F Hilder, Dominic S Peterson, Mingqiang Yi, Frantisek Svec, Jean M J Fréchet, and E O Lawrence Berkeley. 2003. "Fabrication of Porous Polymer Monoliths Covalently Attached to the Walls of Channels in Plastic Microdevices," 3689–93. https://doi.org/10.1002/elps.200305536.

Svec, Frantisek. 2010. "Porous Polymer Monoliths: Amazingly Wide Variety of Techniques Enabling Their Preparation." *Journal of Chromatography A* 1217 (6): 902–24. https://doi.org/10.1016/j.chroma.2009.09.073.

Svec, Frantisek. 2012. "Quest for Organic Polymer-Based Monolithic Columns Affording Enhanced Efficiency in High Performance Liquid Chromatography Separations of Small Molecules in Isocratic Mode." *Journal of Chromatography A* 1228: 250–62. https://doi.org/10.1016/j.chroma.2011.07.019.

Svec, Frantisek, and Christian G. Huber. 2006. "Monolithic Materials: Promises, Challenges, Achievements." *Analytical Chemistry* 78 (7): 2100–2107. https://doi.org/10.1021/ac069383v.

Svec, Frantisek, and Yongqin Lv. 2015. "Advances and Recent Trends in the Field of Monolithic Columns for Chromatography." *Analytical Chemistry* 87 (1): 250–73. https://doi.org/10.1021/ac504059c.

Umemura, Tomonari, Yuji Ueki, Kin Ichi Tsunoda, Akio Katakai, Masao Tamada, and Hiroki Haraguchi. 2006. "Preparation and Characterization of Methacrylate-Based Semi-Micro Monoliths for High-Throughput Bioanalysis." *Analytical and Bioanalytical Chemistry* 386 (3): 566–71. https://doi.org/10.1007/s00216-006-0425-2.

Urban, Jiri, Pavel Jandera, and Pavel Langmaier. 2011. "Effects of Functional Monomers on Retention Behavior of Small and Large Molecules in Monolithic Capillary Columns at Isocratic and Gradient Conditions." *Journal of Separation Science* 34 (16–17): 2054–62. https://doi.org/10.1002/jssc.201100174.

Urban, Jiri, Frantisek Svec, and Jean M J Fréchet. 2010. "Hypercrosslinking: New Approach to Porous Polymer Monolithic Capillary Columns with Large Surface Area for the Highly Efficient Separation of Small Molecules." *Journal of Chromatography A* 1217 (52): 8212–21. https://doi.org/10.1016/j.chroma.2010.10.100.

Wang, Hongxia, Haiyang Zhang, Yongqin Lv, Frantisek Svec, and Tianwei Tan. 2014. "Polymer Monoliths with Chelating Functionalities for Solid Phase Extraction of Metal Ions from Water." *Journal of Chromatography A* 1343: 128–34. https://doi.org/10.1016/j.chroma.2014.03.072.

Wang, Zhuozhuo, Yingmin Liao, Lei Chen, and Xiaojia Huang. 2020. "On-Site Sample Preparation of Trace Aromatic Amines in Environmental Waters with Monolith-Based Multichannel in-Tip Microextraction Apparatus Followed by HPLC Determination." *Talanta* 220 (June): 121423. https://doi.org/10.1016/j.talanta.2020.121423.

Wu, Beibei, Turghun Muhammad, Sailemayi Aihebaier, Kal Karim, Yiting Hu, and Sergey Piletsky. 2020. "A Molecularly Imprinted Polymer Based Monolith Pipette Tip for Solid-Phase Extraction of 2,4-Dichlorophenoxyacetic Acid in an Aqueous Sample." *Analytical Methods* 12 (40): 4913–21. https://doi.org/10.1039/d0ay01587c.

Xie, Shaofeng, Frantisek Svec, and Jean M. J. Fréchet. 1998. "Porous Polymer Monoliths: Preparation of Sorbent Materials with High-Surface Areas and Controlled Surface Chemistry for High-Throughput, Online, Solid-Phase Extraction of Polar Organic Compounds." *Chemistry of Materials* 10 (12): 4072–78. https://doi.org/10.1021/cm9804867.

Xu, Yan, Qing Cao, Frantisek Svec, and Jean M.J. Fréchet. 2010. "Porous Polymer Monolithic Column with Surface-Bound Gold Nanoparticles for the Capture and Separation of Cysteine-Containing Peptides." *Analytical Chemistry* 82 (8): 3352–58. https://doi.org/10.1021/ac1002646.

Zhu, Tao, and Kyung Ho Row. 2012. "Preparation and Applications of Hybrid Organic-Inorganic Monoliths: A Review." *Journal of Separation Science* 35 (10–11): 1294–1302. https://doi.org/10.1002/jssc.201200084.

In: Polypropylene
Editor: Théodore Marleau
ISBN: 978-1-68507-378-7

Chapter 2

Recycled Polypropylene and Their Sustainable Composite of Polypropylene as Novel 3D Printing Materials and Opportunities for the Circular Economy

***A. M. S. Marton*[1], *L. C. S. Balieiro*[1], *N. C. Zanini*[2], *E. J. O. Teixeira*[1], *S. F. Medeiros*[3], *D. S. Rosa*[2] and *D. R. Mulinari*[1*]**

[1]Department of Mechanic and Energy,
State University of Rio de Janeiro (UERJ), Resende, Brazil
[2]Center for Engineering, Modeling, and Applied Social Sciences (CECS), Federal University of ABC (UFABC), Santo André, Brazil
[3]Chemical Engineering Department, Engineering School of Lorena, University of São Paulo, Lorena, SP, Brazil

---

* Corresponding Author's E-mail: daniella.mulinari@fat.uerj.br; dmulinari@hotmail.com.

ABSTRACT

Currently, plastic products have generated severe environmental problems due to the inadequate destination of post-consumer material, which justifies the possibilities of reusing these materials. One of the biggest problems is the large volume that these materials occupy in landfills, and their current low rate of degradation, making their soil complex. Thus, rapid prototyping technology has conquered the market as an innovative technique and, more recently, through using a sustainable source or recycling of polymeric materials, such as polypropylene (PP). Polypropylene is a semi-rigid, lightweight material commonly used in storage and packaging applications. However, the semi-crystalline structure of the material causes great deformation in 3D printed parts on cooling, which is a significant challenge for 3D printing. On the other hand, polypropylene is a strong polymer with excellent fatigue resistance, making it suitable for low-strength applications such as hinges, belts, and leashes. Thus, this chapter aims to show the technical-scientific feasibility of post-consumer polypropylene recycling to obtain 3D printer filaments and present some necessary characteristics of the materials used in FDM technology to ensure final parts with good properties. Furthermore, the use of sustainable polypropylene composite filaments in this area was discussed.

Keywords: recycled plastics, polypropylene, sustainable polypropylene composite, additive manufacturing (AM), 3D Printing

INTRODUCTION

One of society's biggest challenges is to deal with the excessive generation of solid waste and its disposal. More than 9 billion tons of plastic and plastic byproducts end up in the environment annually due to poor waste management. The current waste management and environmental legislations require all engineering materials to be properly recovered and recycled. However, lack of appropriate disposal areas brings consequently global environmental impacts. Environmental policies have guided governments, society, academia, and industry towards sustainable solid waste management. Thus, additive manufacturing (AM), which involves 3D Printing, has encouraged new materials from renewable sources and new manufacturing technologies. The use of recycled plastic material for AM can propose a process that is more environmentally friendly and can solve the problem of plastic waste globally is expected.

Currently, there is also a high demand for personalized, lightweight, and at the same time durable parts that could be manufactured in a fast and flexible, versatile, and low-cost production process. For sustainability reasons, 3D Printing filaments can be manufactured from recycled polymers and composites, reducing materials cost while maintaining their performance and useful life. One of the ways to make recycled plastic filaments is by extrusion. In this extrusion process, many factors can affect the quality of the product.

Among polymeric materials, polypropylene (PP) is a great candidate for products using additive manufacturing. It is low-cost, versatile, has good mechanical and biological properties, chemical resistance, and inertness. These properties result in numerous applications of this material. However, due to the semicrystalline nature of PP, it is not so easy to obtain samples with excellent geometric accuracy as the manufactured components tend to shrink and deform during the AM process. Many researchers have studied ways to improve the PP processability and developed news composites using renewable source materials searching a circular economy. Thus, various reinforcement fibers have been produced by mechanical recycling, thermal recycling, and chemical recycling. For the obtained composites, recycling is needed to develop better recyclable composites and more efficient separation technologies.

In this chapter, the authors report:

- Polymeric materials recycling;
- PP and its recycling;
- Additive manufacturing (AM) technology;
- Recycled PP filaments for use in AM;
- Sustainable PP composite filaments for use in AM;

- PP composite filaments reinforced with other types of reinforcements.

Polymeric Materials Recycling

The global production of plastic grows annually, and this value is related to the insertion of plastic products in different areas of the consumer's daily life, such as packaging, construction, automobiles, electronics, toys, housewares, furniture, footwear, and others (Singh et al., 2017; Dahlbo et al., 2018).

In this context, one of the sectors that most use plastic material, being one of the largest generators of plastic waste in the environment, is the food packaging sector, due to high consumption and short shelf life (Ragaert, Delva and Geem, 2017; Dahlbo et al., 2018). The most found polymers in plastic packaging waste are high-density polyethylene (HDPE), low-density polyethylene (LDPE), polypropylene (PP), and polyethylene terephthalate (PET) (Ragaert, Delva and Geem, 2017).

Thus, an effective and specific alternative for plastic waste would be to intensify the re-use and recycling of these materials to mitigate the impact on nature, instead of being piled up in landfills and the environment (Sanchez et al., 2017).

Besides benefiting the environment by reducing the volume of material to be discarded, the recycling of polymers generates benefits to industries such as material cost reduction and processing. Moreover, every day there is a greater demand to use polymeric materials to replace other materials such as glass, metals, and wood, which generate even more solid waste, making recycling of fundamental importance. Even though plastics provide economic and environmental benefits during their various applications, they have been closely scrutinized for their negative environmental impacts, particularly littering and leakage into terrestrial and aquatic environments (Rajni, 2019). Vast amounts of plastic waste are generated globally, and only a fraction is recycled; most plastic waste goes to landfills. Furthermore, one of the alternatives to minimize this impact is reintroducing this waste back into the market, retaining its economic value, thus preserving natural resources and reducing waste.

Recycling and incineration are standard methods of recovery due to conservation of non-renewable fossil fuels: plastic production uses 8% of the world's oil production, 4% as feedstock and 4% during manufacture; reduced consumption of energy; reduced amounts of solid waste going to landfill can also reduce the municipal solid wastes; and reduced emissions of carbon dioxide (CO_2), nitrogen oxide (NO) and sulfur dioxide (SO_2) (Karmakar, 2020). Incineration presents some problems, such as producing toxic gases and ash residues containing lead and cadmium. Recycling has advantages such as reducing environmental problems and saving material and energy (Grigore, 2017).

Therefore, plastic can be degraded in the environment by four mechanisms: photodegradation, thermo-oxidative degradation, hydrolytic degradation, and biodegradation by microorganisms. The natural degradation of plastic begins with photodegradation due to ultraviolet light from the sun, which provides the activation energy needed to initiate the incorporation of oxygen atoms in the plastic, leading to thermo-oxidative degradation. At this stage, the plastic becomes brittle and breaks into smaller pieces until the polymer chains reach molecular weight low enough to be metabolized by microorganisms. Microorganisms convert carbon from polymer chains into carbon dioxide or incorporate it into biomolecules, but this process can take 50 years. Therefore, recycling can solve these problems, as most commodity plastics are relatively stable, making monomer recovery poor (Grigore, 2017).

When thinking about a linear economy system or circular economy, there are four main types of the recycling process: primary recycling, secondary recycling, tertiary recycling, and quaternary recycling, as shown in Figure 1 (Hopewell, 2009).

These four processes can be used for thermoplastic polymers, but only re-use is feasible for thermosets or using this material as a filler in new recycling systems. Thermoset polymers form chemical crosslinking between chains, meaning that both intermolecular and intramolecular bonds are of the same nature, both covalent, which causes the material to degrade before it can be melted. Thermoplastic polymers form physical bonds between the chains, which makes it possible to fuse several times.

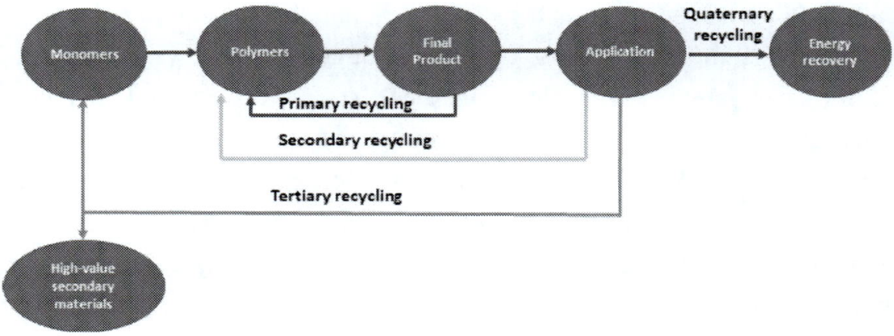

Figure 1. Descriptive illustration of types of plastic polymer recycling.

The primary recycling or mechanical recycling is the direct re-use of uncontaminated discarded polymer into a new product without loss of properties. The manufacturer conducts primary mechanical recycling for post-industrial waste (Salem, Lettieri, and Baeyens, 2010). Therefore, this process is often termed closed-loop recycling. In principle, post-consumer waste can also be subjected to primary recycling; however, in this case, several additional complications may arise, such as the necessity of selective collection (Salem, Lettieri, and Baeyens,2010) and rough (manual) sorting (Baillie et al., 2011). Such issues may significantly increase the costs of recyclates (Salem, Lettieri, and Baeyens, 2010). In general, this method is unpopular among recyclers. Before reintegrating a used material into a new product, it requires typically grinding, shredding, crushing, or milling. These processes make the material more homogeneous and easier to blend with additives and other polymers for further processing. Broken-down material can also be integrated in a more controllable way into a typical production process. Moreover, it becomes easier to purify. An additional cleaning step could be helpful or even necessary to avoid problems that might otherwise occur with the final products (Goodship, 2007).

A recyclate can be given a new shape after melting. The best-known methods of this type of processing of mechanical recyclates are injection molding, extrusion, rotational molding, and heat pressing (Salem, Lettieri, and Baeyens, 2010; Goodship, 2007). Therefore, only thermoplastic polymers, such as polypropylene (PP), polyethylene (PE), polyethylene terephthalate (PET), and polyvinyl chloride (PVC), can commonly be recycled.

The secondary recycling or mechanical recycling, and in primary recycling, only thermoplastic polymers can be reprocessed. The polymer is not changed during the secondary recycling, but its molecular weight falls due to chain scissions, which occur in water and trace amounts of acids. This may result in the reduction of mechanical properties. This phenomenon can be partially counteracted by intensive drying, vacuum degassing, and various stabilizing additives (Achilias, 2012). Another reduction in mechanical properties after recycling is the main polymer (matrix) contamination with other polymers. The waste is ground without prior purification in this recycling, optionally cleaned after grinding, and integrated into the product, mainly through melting. Essential factors of secondary recycling are the availability of waste materials for recycling; costs of collection, storage, and transportation; form or shape; composition; a purity grade; price difference between virgin and recycled materials; the presence of desired and undesired additives; availability and costs of techniques and processes; ecological aspects (Brems, Baeyens, and Dewil, 2012; Goodship, 2007).

Tertiary recycling or chemical recycling is defined as the process in which polymers are chemically converted to monomers or partially depolymerized to oligomers through a chemical reaction (a change occurs to the chemical structure of the polymer). The resulted monomers can be used for new polymerizations to reproduce the original or a related polymeric product. This method can transform the plastic material into smaller molecules, suitable for feedstock material starting with monomers, oligomers, or mixtures of other hydrocarbon compounds (Francis, 2016).

The chemical reactions used to decompose polymers into monomers are hydrogenation, glycolysis, gasification, hydrolysis, pyrolysis, methanolysis, chemical depolymerization, thermal cracking, catalytic cracking and reforming, photodegradation, ultrasound degradation, and degradation in a microwave reactor. Chemical recycling is not fully developed, and for this reason, only a few companies are working on it because this method needs a lot of investment and expert personnel (Francis, 2016).

Quaternary recycling or energy recovery is a method that refers to the recovery of the plastic's energy content. The most effective way to reduce the volume of organic materials, which involves the recovery of energy, is represented by incineration. This method is a good solution because it generates considerable energy from polymers. However, it is not ecologically acceptable because of the health risk from airborne toxic substances, for example, dioxins (in the case of heavy metals, chlorine-containing polymers, toxic carbon, and oxygen-based free radicals).

Among the recycling techniques, the only acceptable according to the principles of sustainable development is chemical recycling. This method leads to forming the monomers from which the polymer is made (Francis, 2016). Also, the advantages and challenges of the techniques are presented in Table 1.

Table 1. Summary of techniques for recycling

Technique	Advantages	Challenges
Mechanical recycling	Cost-effective, efficient, well-known	Deterioration of product's properties, pre-treatment
Chemical recycling	Operational for PET, simple technology	Mainly limited to condensation polymers
Energy recovery	Generates considerable energy from polymers	Not ecologically acceptable

PP and Its Recycling

Polypropylene (PP) is a member homopolymer of the polyolefin family and is widely used. It presents low-density, low-cost thermoplastic semicrystalline. PP is a thermoplastic produced from *propene* (or *propylene*) monomer. The polypropylene chemical formula is $(C_3H_6)_n$, as shown in Figure 2 (Graziano, Jaffer and Sain, 2018).

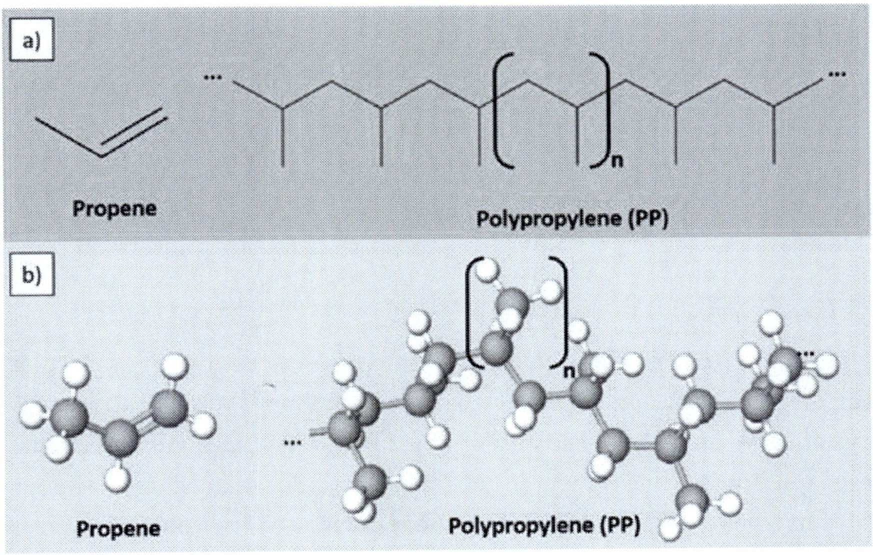

Figure 2. Propene and PP molecular structures in a) 2D and b) 3D arrangement.

The polymerized propene molecule can be formed into three basic chain structures depending on the position of the methyl groups (Figure 3): isotactic (iPP) (a) which the methyl groups (CH_3) arranged on one side of the carbon chain; and syndiotactic (sPP) (b) with an alternating methyl group (CH_3) arrangement and atactic (aPP) (c) with irregular methyl group (CH_3) arrangement (Tsioptsias et al., 2020).

The isotactic arrangement has molecules tightly packed into the crystal structure, which allows the isotactic material to be more rigid and resistant than syndiotactic and atactic (Wang et al., 2018). Consequently, the commercially important PP is isotactic, made by Ziegler-Natta catalysis with catalysts that produce crystallizable polymer chains, a semicrystalline polymer with good physical, mechanical, and thermal properties in the solid state (Maddah, 2016).

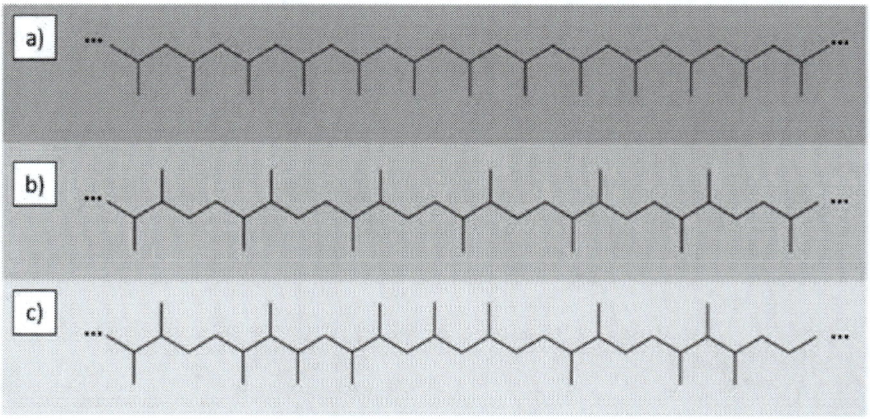

Figure 3. PP and it is polymer chains: a) isotactic, b) syndiotactic, and c) atactic.

The properties of polypropylene depend on tacticity, molecular weight (MW) and molecular weight distribution (MWD), degree of crystallinity, and type (Tzounis et al., 2018). PP offers good chemical resistance in a wide range of chemicals (acids, alkalis, and organic solvents), good fatigue resistance, good resistance to environmental stress cracking, good detergent resistance, and ease of machining, along with the unique ability to be manufactured by any standard processing methods such as injection molding, extrusion, blow molding and thermoforming (Shubhral, Alam, and Quaiyyum, 2011).

PP is among the most used plastics globally, after PE, with numerous industrial and consumer applications, as seen in Figure 4 (Millet et al. 2018).

PP is referred to as commodity and engineering plastic. Examples of numerous PP applications include household goods (bottles, buckets), packaging, automotive industry (bumpers, wheel covers), construction, pipes and fittings, fibers, furniture, as seen in Figure 5 (Spoerk, Holzer, and Gonzalez-Gutierrez, 2019). PP applications are used in different industries such as military, household appliances, automobiles, and construction due to their physical and chemical properties (Kristiawan, 2021). However, PP has low thermal, electrical, and mechanical properties compared to other engineering plastics (PC, PA, etc.) and has a high coefficient of friction under dry shear conditions (Jun et al., 2017).

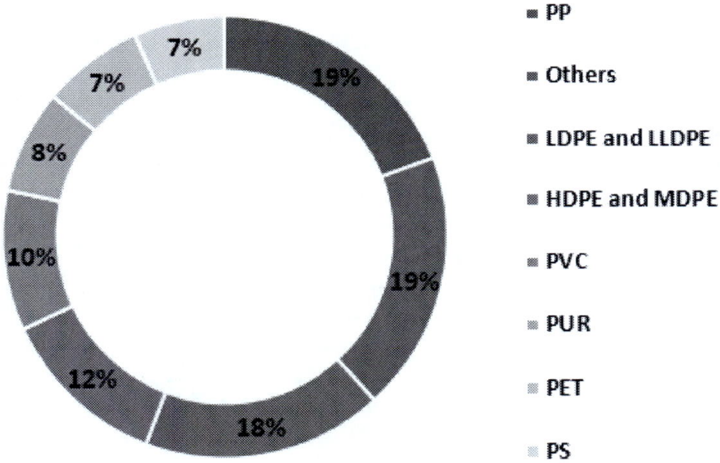

Figure 4. Illustration of the European Plastic Convert Demand by Polymer Type for the year 2017 (Matthews, Moran and Jaiswall, 2021).

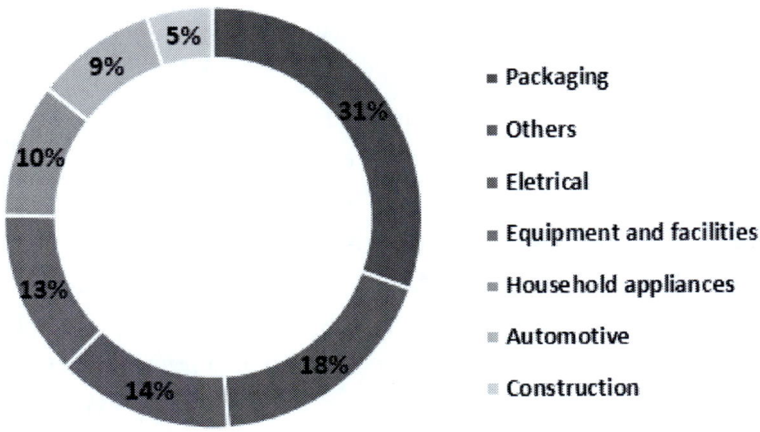

Figure 5. Applications of Polypropylene (Maddah, 2016).

Homopolymers and copolymers are the significant types of PP in the market; they have good mechanical and thermal properties, as shown in Table 2. Homopolymer is the most commonly used type of general-purpose polypropylene. Major applications include packaging, textiles, healthcare, tubing, automotive and electrical applications (Al-Otaibi et al., 2020).

Table 2. Mechanical and thermal properties of polypropylene

Property	Homopolymer			Copolymer	
Melt flow index	3	0.7	0.2	3	0.2
Tensile strength (MPa)	34	30	29	29	25
Elongation at break (%)	350	115	175	40	240
Flexural modulus (MPa)	1310	1170	1100	1290	1030
Brittleness temp. (°C)	+15	0	0	-15	-20
Vicat softening point (°C)	154-150	148	148	148	147
Rockwell hardness (R-scale)	95	90	90	95	88.5
Impact strenght (ft lb)	10	25	34	34	42.5

Fonte: Maddah, 2016.

Some characteristics differentiate the homopolymer from the copolymer. The homopolymer is more resistant than the copolymer, can contact food, has good chemical and welding resistance, and has good processing and impact resistance. While the copolymer has good impact resistance, is soft, rugged, and durable, withstands low temperatures, and is unsuitable for food contact (Al-Otaibi et al., 2020).

The family of copolymers is further divided into random copolymers and block copolymers produced by the polymerization of propene and ethane: random copolymer is produced by polymerizing ethylene and propene. It has ethylene units, usually up to 6% by mass, randomly incorporated into polypropylene chains. These polymers are flexible and optically clear, making them suitable for applications requiring transparency and products that require an excellent appearance (Lai and Men 2018). While in block copolymer, the ethene content is higher (between 5 and 15%). In addition, it has comonomer units arranged in a regular pattern (or blocks). The regular pattern thus makes the thermoplastic stronger and less brittle than the random copolymer. Thus, these polymers are suitable for applications that require high strength, such as industrial uses (Feng et al., 2017).

Impact copolymer is functional on parts that require good impact resistance. Impact copolymers are used primarily in packaging, household appliances, film, tube applications, and the automotive and electrical segments (Maddah, 2016).

Expanded polypropylene is an ultra-low density, closed-cell sphere foam. EPP ball foam has increased weight resistance, excellent impact resistance, thermal insulation, and chemical and water resistance.

Polypropylene terpolymer comprises propylene segments joined by ethylene and butane monomers that appear randomly along the polymer chain. PP terpolymer has better transparency and makes it suitable for sealing film applications.

High melt strength Polypropylene (HMS PP) is a branched long chain material that combines high melt strength and melt phase extensibility. HMS PP is used to produce low-density soft foams for food packaging applications and is used in the automotive and construction industries; it has various mechanical properties, high thermal stability, and good chemical resistance.

Since PP is fatigue resistant and durable material, it has been an excellent candidate to fabricate products by additive manufacturing (AM) techniques (Wang et al., 2019; Carneiro, Silva and Gomes, 2015).

Recycled Polypropylene (rPP)

The consumption of plastics for products and packaging continues to grow, despite environmental and sustainability concerns. More efficient reprocessing techniques and value-added applications are likely to be needed to change these recycling trends (Zander et al., 2019). Recycling is essential to reduce the number of materials sent to landfills and the use of raw materials. Direct use of post-consumer polymers is the most efficient way to treat plastic waste (Pinho, Amaro and Piedade, 2020).

The biggest problem is plastics made from HDPE, LDPE, PP, and PVC, used mainly by manufacturers and landfilled with greenhouse gas emissions (Mikula et al., 2021).

Figure 6. Resin identification code for PP.

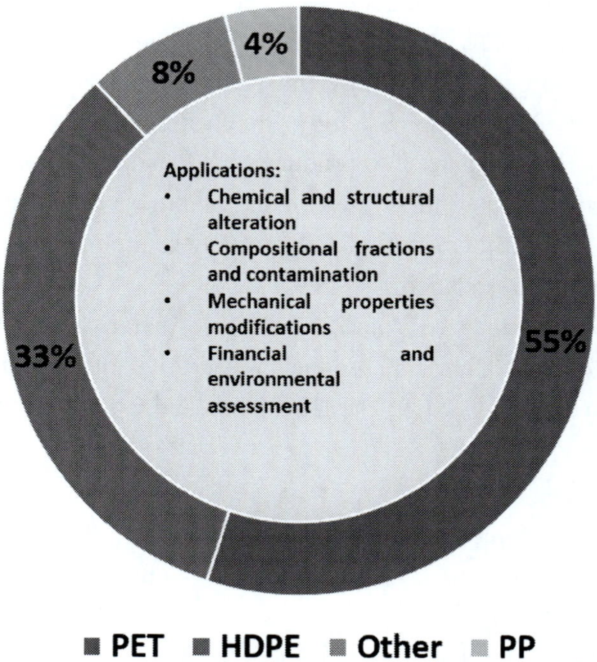

Figure 7. Global Recycling Market (Vidakis et al., 2021).

PP is among the three most used plastics globally, after PE and PVC, with many industrial and consumer applications, and represents one of the primary sources of plastic waste. Therefore, the first step in the PP recycling process is to ensure that it is separated from any other plastic polymers, which can be easily recognized using the resin identification coding system established by the Society of the Plastic Industry (SPI) and is printed on the bottom of a product (Figure 6).

Despite the low price of polypropylene, companies prefer to use virgin materials instead of recycled ones. Therefore, many studies have proposed the use of post-consumer PP for 3D Printing. Thus, it can add more value to recycled material and offer the market a material that can contribute to the environment (Pickering and Stoop, 2017). Currently, the overall recycling market reaches 72.6 billion dollars, while 4.0% belongs to the PP with an estimated value of 2.9 billion dollars (Vidakis et al., 2021), as shown in Figure 7.

However, some more significant difficulties in recycling homogeneous PP arise from the easy degradability of this polymer during processing and recycling operations. Polypropylene is eventually affected by thermal degradation; the bonds between hydrogen and carbon weaken. Furthermore, mechanical stress and ultraviolet radiation also strongly alter the structure and morphology of PP. Elongation at break and impact strength are the properties most influenced by degradation. As PP has tertiary carbon in the chain, the impact of thermal and mechanical degradation on PP is even more dramatic than on other polymeric materials (Iunolainen, 2017).

Table 3. Material properties of isotactic PP, ABS, and PLA (Iunolainen, 2017)

Material Properties	PP	ABS	PLA
Density (g.cm^{-3})	0.90-0.91	1.0-1.4	1.3
Glass Transition Temperature (°C)	-20-20	-85-105	60-65
Melting Point (°C)	160-165	Amorphous	150-160
Processing Temperature (°C)	200-230	210-230	160-220
Heat deflection temperature (HDT) at 0.46 MPa (°C)	100	98	50
Tensile Strength at yield (MPa)	26	34	60
Young´s Modulus (GPa)	1.3-1.5	1.7-2.8	2-3.5
Flexural Modulus (GPa)	1.5-2	1.5-2.7	2.4-4
Impact Strength (J.m^{-1})	150	250	47

Thus, the application of PP in 3D Printing is challenging to print due to severe warping and poor layer adhesion; only very few printers can print with PP now (Iunolainen, 2017). The specific physical and mechanical properties of unmodified PP are compared to the two most popular materials for 3D printer filaments, ABS and PLA, as shown in Table 3.

Analyzing Table 3, in terms of stiffness, it is noted that PLA has the highest values. Moreover, as PP has a low flexural modulus due to its elongation capability, it is considerably more elastic than PLA. Because of this, it is problematic to print strong hinges for lids with PLA, which is fragile and always tends to break rather than bend.

Using PP for 3D Printing is excellent for applications that need to be lightweight, waterproof, or durable. Food or liquid containers and clips and latches that fold repeatedly are popular uses. According to Craftcloud (2021), Printing PP can be a challenge, but it can increase success with the correct print settings. PP is challenging for its tendency to come off the print bed.

While not as commonly available as filaments such as PLA, ABS, or PETG, polypropylene is slowly expanding its variety and availability in the market and becoming more affordable for consumers.

Additive Manufacturing (AM) Technology

Additive manufacturing (AM) is a tool of industry 4.0, popularly known as 3D Printing. It adds material rather than removes it and can produce three-dimensional objects close to or equal to the original dimensions (Tofail et al., 2018). AM processes do not require molds or special tools to make complex shapes (Kristiawan et al., 2021). As is apparent from its very name, AM, popularly known as 3D Printing, adds material rather than removes it and can produce three-dimensional objects close to or equal to the original dimensions (Tofail et al., 2018). The process of 3D Printing happens through the melting and posterior solidification of materials by layer-by-layer depositions, gradually forming the geometry of the desired product without the necessity for significant finishing or wasting material (Tofail et al., 2018; Zhang et al., 2019).

Besides the benefit of geometric freedom of computer-aided design software and scanners, providing engineers to materialize virtual concepts (Kelly et al., 2019), the use of AM has been one of the most sought-after routes in the contemporary world. The global AM market is expected to reach a compound annual growth rate of 14.4% by 2027, stimulated by end-use and customization industries. Thermoplastic polymers can be the most widely used in the forecasted period between 2020 and 2017 due to their cost benefits for industrial and home purposes. In North America, the United States is at the forefront of the use of AM. However, the European continent stands out for the big automotive companies (Mercedes and Volkswagen) and the aerospace segment (Airbus) in aerodynamic components and complex geometries, made possible by cutting-edge AM technologies. Asia-Pacific is another emerging area globally, with AM investment forecast to be close to four billion dollars by 2026. (Reports and Data, 2021).

The continuous growth and search for AM are due to sustainable concerns regarding reduced energy and input consumption, clean production, and environmentally friendly methods (Majeed et al., 2021). Another factor that can further increase the green character of AM can be the re-use of materials. Thus, from an industrial perspective, in addition to resignifying the waste for a new purpose, the AM can be essential for a transition from a linear economy to a circular economy (Cruz Sanchez et al., 2020). In terms of availability and quantity, recyclable waste plastics and vegetable biomass are suggested for re-use as materials for applications in AM. However, specific 3D Printing technologies should be used for this purpose, suitable for processing polymers, polymer blends, and polymer composites.

Simplistically, polymer-based 3D Printing technologies are chosen based on two factors. First, the print method depends on the initial state of the polymeric material (whether it is in the form of liquid polymer resin, polymer powder, polymeric filament, or polymeric film).

Table 4. Polymer-based AM technologies regarding their respective polymer initial states, the pattern or fuse used for 3D Printing, and their brief description

The initial state of the polymer	Name/Acronym of the AM technique	Pattern or fuse	Brief description of the AM technology to form the 3D object
Filament	FFF/FDM	Thermal	A heated nozzle that deposits a molten thermoplastic polymer filament and solidified in layers through a moving printhead on pre-set axes (Heidari-Rarani, Rafiee-Afarani and Zahedi, 2019)
Films	LOM	Thermal	Thermoplastic polymer films or sheets are bonded together (or cut) layer-by-layer (Yaragatti and Patnaik, 2020)
Powder	SLS	Optical	Powder bed fusion (PBD) based process, where a laser hits the polymer powder, melting it layer-by-layer (Chen et al., 2021)
	SHS	Chemical/ Mechanical	Same principle as SLS, except that it uses a thermal printhead instead of a laser (Rajamani and Balasubramanian, 2019)
	Binder Jetting	Chemical/ Mechanical	A binder connects the exposed particles of a powder bed without a heat source, layer-by-layer (Ziaee and Crane, 2019)
Liquid resin	SLA	Optical	Photopolymer resin is cured by UV light, forming the object's shape upside down in the resin pool (Stansbury and Idacavage, 2016).
	MultiJet/PolyJet	Optical	A similar process to SLA. Nonetheless, it uses an inkjet-like print head to distribute the resin onto the print bed (Nowacki and Sieczkiewicz, 2020)
	DLP	Optical	A similar process to SLA. However, it is faster due to a projection into the resin tank of an entire surface to be cured at once (Kadry et al., 2019)

However, the choice also depends on the building method used, i.e., the machine pattern or fuse (optical, chemical/mechanical, or thermal). Therefore, polymer-based AM addressed in the study by Stansbury and

Idacavage (2016) can be exemplified as filament fabrication (FFF)/fused deposition molding (FDM), stereolithography (SLA), selective laser sintering (SLS), selective heat sintering (SHS), digital light projection (DLP), multiJet or polyJet, binder jetting, and laminated object manufacturing (LOM) (Stansbury and Idacavage, 2016). For each of the mentioned AM technologies, Table 4 provides information about their characteristics.

Among the polymer-based AM technologies mentioned in Table 5, FDM is the most popularly known, more than the first invented 3D Printing technique (SLA) Wickramasinghe, Do and Tran, 2020). FDM was the AM technology that gathered the maximum 2019 market share, and its growth is expected to be driven by the automotive and healthcare products industries. The key global manufacturers in the FDM printer market are Stratasys Ltd., Exone, XYZprinting, Inc., Mcor Technologies Ltd., Materialise NV, Ultimaker, Optomec, Ricoh Company, Ltd., Proto labs, INC., Organovo Holdings, Inc., Voxeljet AG, EnvisionTEC, Prodways, EOS GmbH, 3D Systems (MarketWatch, 2021). FDM technology is not restricted to research in materials science or industry for new product development and rapid prototyping (Vardhan, Kumar and Chohan, 2019). Nowadays, there are abundant open-source FDM printers due to the expiry of patents. That fact caused a market revolution, reducing the cost of an entry-level printer under a thousand dollars and as low as five thousand dollars (Stansbury and Idacavage, 2016). The economic benefit allows greater accessibility for personal use desktop-style printers, stimulating the maker culture and do-it-yourself (DIY) (Camburn and Wood, 2018), enabling a future scenario that might be a technology increasingly present in everyday human life.

The FDM 3D printing process needs to go through steps prior to the Printing itself. The desired shape is obtained from CAD software, for example, AutoCAD or SolidWorks. After this, the digital drawing must be transformed into a standard triangle language (STL) (Persad and Mencia, 2020; Singh, Chawla and Singh, 2020). Formerly, the shape is transformed into slices (Slic3r software) that, when overlapped, acquire the desired 3D virtual shape. Only after these preliminary steps, the FDM printer goes into action (Figure 8). In the first step, the printer is fed a thermoplastic filament that must have a suitable diameter. Usually, the most conventional commercial filaments are 1.75 mm or 3 mm (Mirón et al., 2017). There is a plethora of thermoplastic filament materials for FDM printing that might be covered in future topics in this same chapter. After inserting the filament in the machine, the thermoplastic material melts to acquire a certain fluidity, respecting the melting and degradation temperature of the polymer in question. After this melting process, the material is extruded by a warm nozzle of the movable heated printhead (x and y-axis), which deposits the delimited layers on a working bed (z-axis) Singh, Chawla and Singh, 2020; Vardhan, Kumar and Chohan, 2019). The filaments' thermoplasticity causes them to solidify at room temperature in their slices, layer-by-layer, materializing the three-dimensional shape (Kristiawan et al., 2021).

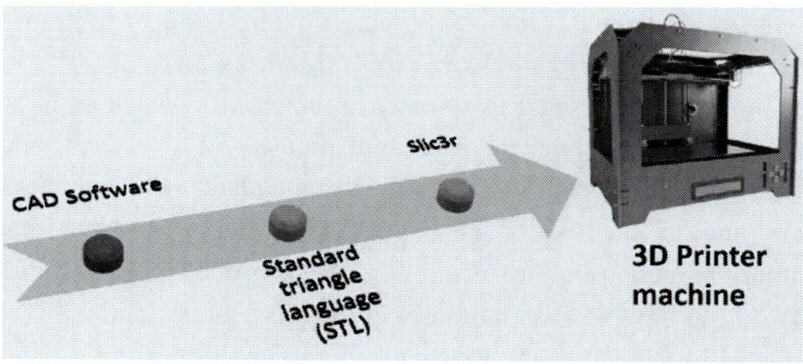

Figure 8. Preliminary stages before the FDM printing.

According to Singh, Chawla and Singh (2020), while more complex techniques like SLA can have brittle printed end-products with fractures at low loads, FDM printed materials can have superior mechanical performance, being able to withstand stresses in the 29-124 MPa range (Singh, Chawla and Singh, 2020). Despite being one of the most assessable AM, FDM depends on several parameters that affect the print quality (Kristiawan et al., 2021). First, the filament feeding the FDM printer must be heated to a temperature that allows the polymeric material to flow for easy extrusion through the nozzle.

Table 5. Mechanical performance parameters divided by FDM machine parameters and the working parameters

Main mechanical performance parameters	Subdivisions	Description	Effect in the 3D printing process
Working parameters	Raster angle	It corresponds to the angle formed between the path taken by the nozzle and the x-axis, which can be between 0 and 90° (Wu et al., 2017)	Influences the forming accuracy and mechanical strength of the printed object since it is related to the loading direction of stress (Wu et al., 2017, 2015)
	Raster width	It corresponds to the width of the layer of material extruded and deposited, forming a pattern. It depends on the nozzle tip (Gupta and Bit, 2018)	When the raster width is large, the interior of the product is more robust. When the raster width is thinner, Printing takes less time and less material (Kristiawan et al., 2021)
	Build orientations	Setting the x, y, and z axes orientations (Kristiawan et al., 2021)	Determine the end-use product surface quality (Kristiawan et al., 2021)
FDM machine parameters	Bed temperature	The working bed where the material is deposited can be heated or not, depending on the polymer used (Aliheidari et al., 2017)	A heated bed can reduce defects of the printed object, such as shrinkage (Choi et al., 2016)
	Nozzle temperature	The movable printhead nozzle where the print layers are extruded has adjustable temperature control (Alsoufi and Elsayed, 2017)	Can affect the warping effect and the dimensional accuracy of the printed object (Akbaş et al., 2020; Alsoufi and Elsayed, 2017)
	Nozzle diameter	The diameter of the nozzle controls the amount of melted polymer that passes through the orifice and be deposited on the working bed	It can affect the thickness of the layers deposited, the surface finish of the printed object (roughness), and also the tensile strength supported by the object (Kiński and Pietkiewicz, 2020; Kristiawan et al., 2021)

Once it reaches the working bed, the filament must solidify to receive a new layer soon. The ideal would be to use a thermoplastic filament with a low coefficient of thermal expansion, low melting temperature, and low glass transition temperature. The aforementioned characteristics prevent print defects such as cracking and warping (shrinkage of the printed object from the edges in contact with the working bed) (Jafferson and Chatterjee, 2021). The mechanical performance parameters are also important. The main ones can be divided between working parameters (usually set in the slicing step with software before the 3D Printing itself) and the FDM machine parameters (Kristiawan et al., 2021). The characteristics of both parameters and their influence on Printing can be seen in Table 9.

The above parameters need to be investigated by the operator of the FDM printer to achieve successful Printing. However, there is a more versatile variation of the FDM technique. So-called 3D pens are considered a handheld FDM and are becoming popular due to their numerous facilities. Compared to conventional FDM, the 3D pen has advantages such as low cost (about 10 dollars), small dimensions (similar to a conventional pen making it easy to handle and move around), no prior CAD drawing is required (the shapes are drawn by hand), and uses the same thermoplastic filaments as used in FDM printers. The operation of a 3D pen is a facilitated version of FDM, where a filament is inserted into the pen, and when the button is pressed, the movement of the filament inside is activated, as well as the heating of the filament, which then passes through the orifice, enabling the 3D pen operator to draw the desired shape with freedom and practicality (Cardoso et al., 2020; De Araujo et al., 2020).

The thermoplastic filament material can affect FDM printing and the 3D pen technique. Pure polymers usually have characteristics that their manufacturers tabulate. However, materials such as blends, recycled polymers, and polymers with added particulate or fiber fillers (polymer composites) may have more differentiated characteristics that should be investigated thoroughly.

Recycled Polypropylene Filaments for 3D Printing

The demand for a new and sustainable alternative material for 3D printer filaments has increased. An alternative source for filament is to re-use and recycle post-consumer plastic products, which is also an attractive approach to conserving energy and sustaining the environment. Furthermore, 3D printing is considered a clean processing technology, as using post-consumer recycled polymers to produce new components/parts can use the circular economy to contribute to the global plastic contamination problem (Pinho, Amaro and Piedade, 2020).

The manufacture of polymeric filaments for the 3D printer is usually done through the extrusion process. The starting material can be a powdered or granulated thermoplastic resin or a recycled post-consumer product fed into the extruder through the feed funnel and converted into a uniform, continuous melt. The melted plastic is molded through a matrix and finally extruded into its final form using standardized diameters that are compatible with the input of the 3D printer (Exconde et al., 2019).

The recycling process of polymeric materials for 3D printing is based on several activities such as selective material separation, decontamination, purification, grinding, re-melting, and extrusion. The main obstacles resulting from this process are logistical and economic aspects. However, given the increasing environmental restrictions and plastic waste recycling, this could be a potential solution, despite the lack of clear economic profitability (Mikula et al., 2021).

The most well-known plastics used in 3D printing (thermoplastics) and their popular applications are shown in Table 6 (Singh et al., 2020).

Amorphous thermoplastic polymers are the perfect materials that can be applied in this process because of their low thermal expansion coefficient, low melting temperature, and less glass transition temperature, properties that can reduce internal stress that includes warping caused during cooling (Jafferson and Chatterje, 2021).

Some recycled plastic materials for 3D Printing became recently available, such as recycled Acrylonitrile Butadiene Styrene (rABS), Polypropylene (rPP), Polyethylene Terephthalate (rPET) or Polylactic Acid (rPLA) (Rigon; Ricotta; Meneghetti, 2020; Zander et al. (2019)). Figure 9 evidence a general scheme of waste recycling for 3D printing.

Table 6. Thermoplastic polymers in 3D printing with their applications

Thermoplastic	Applications
Polyethylene Terephthalate (PET)	Packaging films, PET bottles, staple fiber, carpet yarn
Polypropylene (PP)	Mats, carpet, and rugs
Polyamide (PA6)	Brushers, cams, tires, bearings, gears, threads, ropes
Polycarbonate (PC)	Electrical and Telecommunications Hardware
Polylactic acid (PLA)	Musical instruments, Automotive trim components Drainpipes
Polyvinyl chloride (PVC)	Optical disc drivers, Packaging of computer hardware's Trays, Screen cards
High-density polyethylene (HDPE)	Utensils, toys, pipes, films, bottles, wires, and cable insulations
Polystyrene (PS)	Smoke Detector Housing, Dinnerware, CD cases

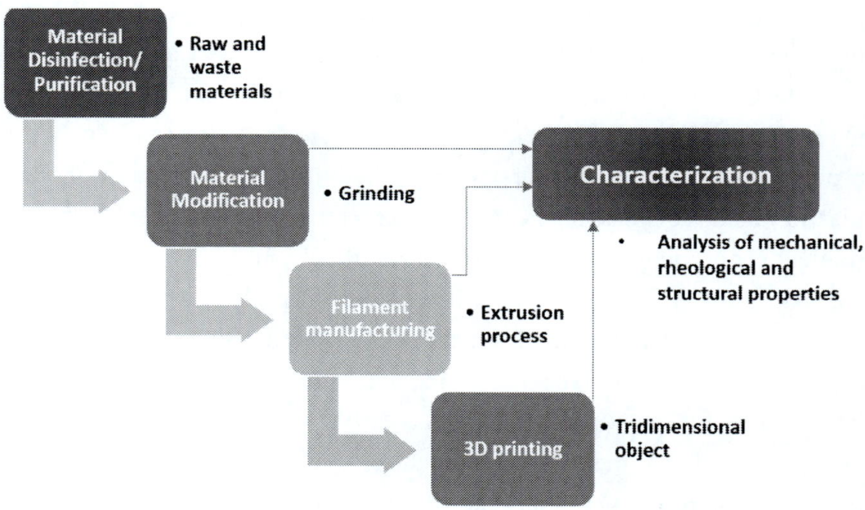

Figure 9. Schematic illustration of the use of plastic residues in the manufacture of filament.

Firstly, the material is segregated and washed, and then the plastic is ground. After, the ground material is extruded at a high temperature, and the prepared filament is inserted into the 3D printer. The printed element is subjected to analysis (mechanical, rheological, and structural properties) (Mikula et al., 2021).

The literature presents few works with recycled PP for use in 3D printer filaments. Iunolainen (2017) investigated the feasibility of manufacturing good quality 3D printer filament from provided CIRCO® recycled PP to find optimal extrusion parameters for manufacturing rPP filament. After parameters optimization during filament production, the ultimate result was still unsatisfactory in terms of quality. The produced rPP filament has quite large diameter variations and is not perfectly round-shaped but rather elliptical. MFI values comparison of the produced rPP filament with the two prevalent commercial filaments ABS and PLA was also made to see how the materials correlate on this score. The recycled PP showed considerably higher flow than both ABS and PLA, and at least for this parameter, CIRCO® rPP is not comparable to them.

Zander et al. (2019a) studied blends of waste PET, PP, and polystyrene (PS) processed into filaments for 3D Printing. The effect of blend composition and styrene ethylene butylene styrene (SEBS) compatibilizer on the resulting mechanical and thermal properties were probed. Recycled PET had the highest tensile strength at 35 ± 8 MPa. Blends of PP/PET compatibilized with SEBS, and maleic anhydride functionalized SEBS had tensile strengths of 23 ± 1 MPa and 24 ± 1 MPa. The non-compatibilized PP/PS blend had a tensile strength of 22 ± 1 MPa. PP/PS blends exhibited reduced tensile strength to ca. 19 ± 1–3 MPa with the addition of SEBS. Elongation to failure was generally improved for the blended materials compared to neat recycled PET and PS. The glass transition was shifted to higher temperatures for all blends except the 50–50 wt. % PP/PET blend. Crystallinity was decreased for the blends but was highest in the 75–25 wt. % PP/PS and the 50-50 wt.% PP/PET blend with SEBS-maleic anhydride. Solvent extraction of the dispersed phase revealed polypropylene was the matrix phase in both the 50–50 wt. % PP/PET and PP/PS blends.

Wang et al., (2019) investigated the thermal properties of polypropylene block copolymer (PPco) and spray-dried CNF (SDCNF)-PPco composite pellet formulations in extrusion-based AM. Thermogravimetric analysis data showed that the addition of SDCNF lowered the thermal degradation onset temperature and increased the residual mass content (at 450°C) of PPco pellets.

Table 7. List of recycled and virgin PP for 3D printer commercial filaments

Material	Additives	Origin
PP	---	McDonnough plastics, InC
PP		Leapfrog 3D printers
rPP	---	Craftcloud® by All3DP
PP	Hemp fiber Harakeke fiber MAPP (maleated polypropylene) Recycled gypsum	Granules of pre-consumer recycled PP supplied by Transpacific Recycling Centre.
PP	Ultramaker PP	Ultramaker Ltd.
PP	BCN3D	Basf SE
PP	Centaur	Form Futura BV
PP	Dddrop	Dddprod BV
PP	P-Lene	TreeD- SA2P di Dario Negrelli Pizzigoni
PP	Ultrafuse PP GF30	Basf SE
PP	FL100PP	2m3D Indústria e Comércio de Plásticos LTDA

Atsani and Mastrisiswadi (2020) optimized the extrusion process parameters to manufacture recycled plastic filaments (polypropylene) that have never been conducted before. The extrusion process was analyzed using the Taguchi and ANOVA methods. This study indicates that the spooler speed of 4 rpm, extrusion speed of 40 rpm, and extrusion temperature of 200 °C in the parameter setting produces an average filament diameter of 1.6 mm. However, filaments made from recycled plastic (polypropylene) have a rough and easily curved surface, so further research is needed.

Vidakis et al., (2021) evaluated the recyclability of Fused Filament Fabrication (FFF) 3D printed polypropylene (PP) for a certain number of recycling courses (six in total), and its effect on the mechanical properties of 3D printed parts. Results showed a slight increase in the material's mechanical properties, both in tension and flexion mode. At the same time, the DSC characterization indicates an increase in the polymer crystallinity over the recycling course.

Table 7 lists the main types of recycled PP and virgin PP for 3D printing commercial filaments.

Recycled Polypropylene (rPP) Filament for 3D Printer: Morphological and Thermal Properties

Due to the few works found in the literature, we propose making a recycled PP filament from plastic packaging for use in 3D printing filament. First, the plastic packages were collected, cut, washed, and dried at 80 °C. After drying, the material was mixed in a thermokinetic mixer (MH-50H, 48 A, MH Equipamentos Ltda., Guarulhos, SP, Brazil) at high rotation speed (5.250 rpm) used to process 70 g in 60 s. After room cooling (60 min), the samples were milled in a knife mill (Plastimax brand, 3.7 kW, Rio Grande do Sul, Brazil) and oven-dried for four hours at 50°C. A mini extruder (Weellzoom brand, model B Desktop, Guangdong Prov, China) was fed with the ground samples to obtain the PPr filament. The parameters used for filament extrusion were extrusion speed to obtain filaments of 85 ± 1 $mm.min^{-1}$ and extrusion processing temperature of 170 ± 0.5°C. Figure 10 evidences the main steps of the rPP filaments processing.

After, filament diameter was measured (taken through a digital caliper, brand Mitutoyo). The filament apparent density was measured using the weight obtained by an analytical balance and an average between 3 diameter measurements of different parts of the extruded filaments (taken through a digital caliper, brand Mitutoyo). The rPP filaments were tested in the printing 3D pen (3DPEN-2, China). The morphology of the rPP filament was analyzed using an optical microscope (brand ZEISS, model Axio Imager 2, New York, USA). The images had magnifications at 0.8x (5mm scale) and 2.5x (1mm scale). The DSC measurements of rPP filaments were performed in TA Instruments (USA) Q20 equipment. The sample (about 6 mg) was heated at a rate of 10°C min^{-1}. The first ramp was performed from −80°C to 180°C to remove possible interference and impurities and the second one from −80°C to 250°C. A quick cooling was carried out using liquid nitrogen to determine the crystallization temperature of the materials.

A series of test trials were done before obtaining the good quality filament from recycled PP during the filament extrusion. Extruded filament can be observed in Figure 11. The measurement of the filament diameter was 1.73 ± 0.05 mm, similar to the rPP filament studied by Iunolainen (2017).

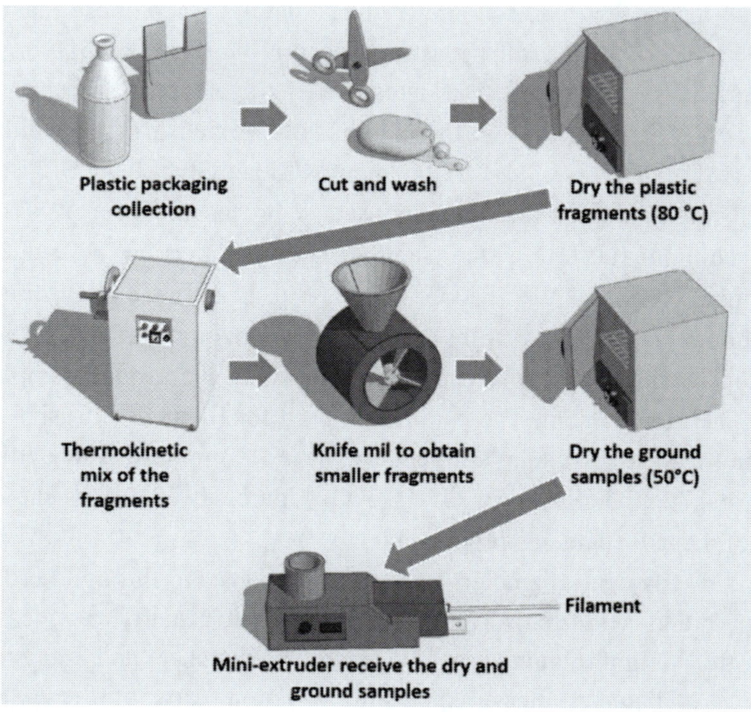

Figure 10. Schematic illustration of the adopted processing methodology of the rPP filament.

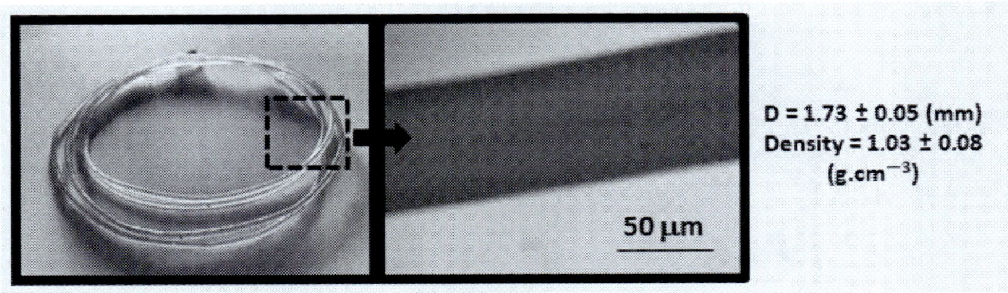

Figure 11. Extruded rPP 3D filament for a 3D pen, optical photomicrograph, diameter, and density.

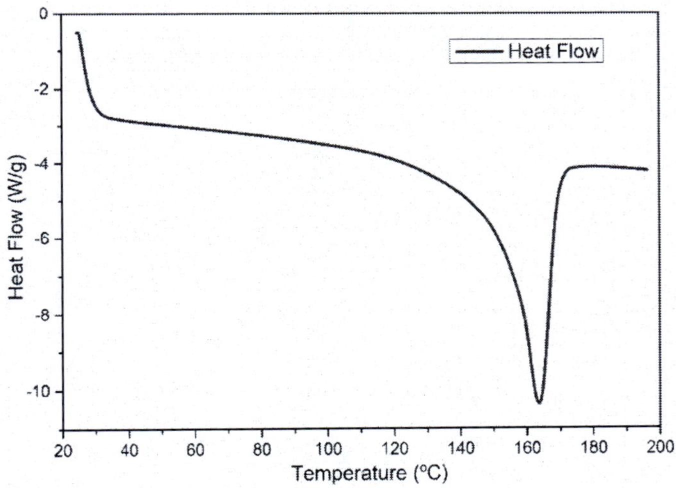

Figure 12. DSC curve of second heating for rPP filament.

Figure 12 presents the thermogram of a rPP filament sample obtained in differential scanning calorimetry (DSC) equipment (TA Instruments, USA). The samples were heated at a rate of 10°C·min^{-1}, from 25 to 200°C under a nitrogen atmosphere. Thermograms were obtained by the second heating cycle to eliminate the impurities in the sample. rPP filament thermogram showed a melting point at 164°C. The crystallinity degree was determined by the equation X_c (%) = ($\Delta H_m / \Delta H^0{}_m$), where $\Delta H^0{}_m$ is the fusion enthalpy of 100% crystalline PP (171 J.g-^{1}) (Lanyi et al., 2019) and ΔH_m is measured by DCS (69.05 J.g^{-1}). The crystallinity rate of the rPP filament was 40%.

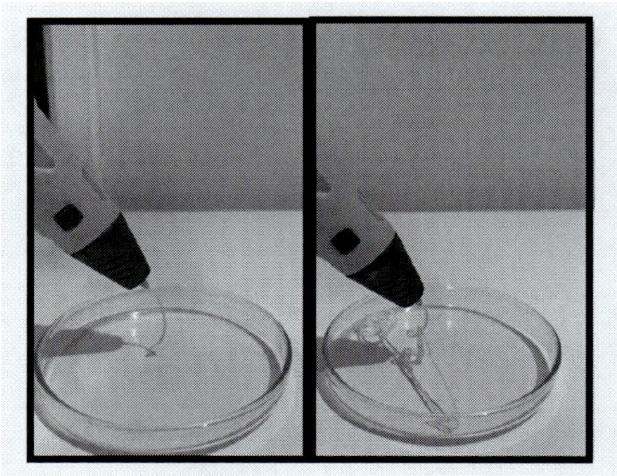

Figure 13. Printing 3D pen of rPP filaments.

Filaments were tested in the printing 3D pen and presented satisfactory results (regarding visual appearance, shape, and stability). Figure 13 evidences 3D pen printable filaments rPP.

Sustainable Composites of Polypropylene Filaments for 3D Printing

In recent years, three-dimensional (3D) printers and low-cost extruders have been expanding to manufacture filament and the use of recycled plastic material for 3D printing has increased. Plastics recycled from virgin materials such as faulty prints can usually be recycled at least once without mechanical properties. However, most polymer packaging materials, such as polyolefins, need some reinforcement to obtain properties similar to common commercial materials used in manufacturing polymer additives (Zander et al., 2019).

The addition of reinforcing fibers aims to improve mechanical properties, adding value to recycled polymers and potentially opening needed markets for the material, and mitigating the environmental impact

(Immonen et al., 2021; Mohammady and HKhairou, 2021; Shulga et al., 2020; Stoof and Pickering, 2017).

Thus, composites materials can provide more sustainable material solutions. For industry, sustainability means reducing material and energy consumption, using renewable material and energy sources, recycling, and minimizing or eliminating environmental impact from emissions and hazardous chemicals. Additive manufacturing methods have several sustainability benefits and can create complex geometries difficult or impossible to produce with conventional methods (McCarthy and Brabazon, 2021; Ford and Despeisse, 2016).

It has recently been shown that the addition of natural fibers to rPP improves its mechanical properties, thermal resistance, and flowability and improves its eco-friendly credentials (Ahmed et al., 2020). Other renewable resources can also be added to rPP, such as marine biomass, wood, and agricultural residues, which are potential alternatives to fossil resources. These resources can minimize the generation of residues and maintain the value of products, materials for as long as possible, contributing to the bioeconomy (Stegmann; Londo; Junginger, 2020).

As an alternative to producing a natural fiber-reinforced thermoplastic filament, mixing can be achieved using a co-rotating twin-screw extruder or dryer, allowing for dispersive and distributive mixing (Costa et al., 2021; Ahmed et al., 2020). Even though fiber reinforcement appears feasible and promising, it has various challenges that need to be overcome, such as the effect of fibers on the resolution, agglomerate formation, the heterogeneous composite formation, blockage of printer heads, non-adhesion, and increased curing times (Ahmed et al., 2020).

When using natural fibers, dry them carefully in the initial phase, even before the composition. It is essential to reduce the water content, leading to hydrolytic degradation, poor quality filaments, or unsatisfactory results. The temperature used must be carefully monitored to avoid thermo-oxidative degradation. If the viscosity of the material increases, it is crucial to have high extrusion temperatures (Mazzanti; Malagutti; Mollica, 2019).

It is also important to mention that the type and size of the nozzle must be chosen carefully. Too narrow a tip can lead to sieving of the filament during extrusion, which can lead to uneven material flow and introduce defects in the printed sample. Likewise, a wide nozzle can release more than the required amount of molten matter into the plate, making accurate shape difficult, resulting in a very warped and uneven printed sample (Ahmed et al., 2020).

On the other hand, natural fibers can decrease the environmental impact of the composite, obtain lower density, and are less stiff compared to traditional inorganic fillers. Because of this, natural fiber-reinforced polymers, commonly known as biocomposites, are gaining essential market shares in many industrial sectors such as automotive, construction, thermally insulating, and sound-absorbing materials (Mazzanti; Malagutti; Mollica, 2019). In FDM, the use of thermoplastic biocomposite filaments is interesting for decreasing material cost and environmental impact, reducing distortion after processing while possibly preserving the material's mechanical properties (Gui et al., 2018).

Figure 13 illustrates the basic steps of 3D Printing using composites consisting of polymers and natural fibers.

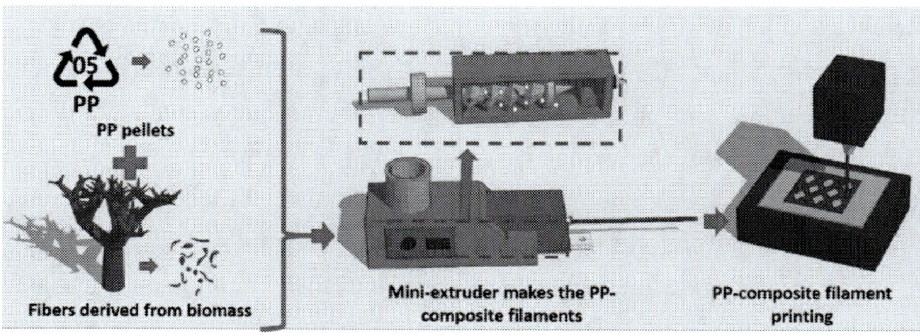

Figure 13. Steps involved in the 3D printing process using composites consisting of polymers and natural fibers.

Milosevic, Stoof and Pickering (2017) investigated and characterized the performance of natural fiber reinforced polypropylene composites in fused deposition modeling (FDM). Composite filaments comprising pre-consumer recycled polypropylene with varying hemp or harakeke fibers were extruded from which tensile test specimens were made using FDM. Filament and test specimens were tensile tested, and properties were compared with plain polypropylene samples; the ultimate tensile strength and Young's modulus of reinforced filament increased by more than 50% and 143%, respectively, for both 30wt % hemp or harakeke compared to polypropylene filament. Results showed to give benefit regarding dimensional stability during extrusion and FDM.

Stoof and Pickering (2017) described the preparation of composite filaments with differing harakeke, hemp fiber, or recycled gypsum contents (0-50wt%) in pre-consumer recycled polypropylene (PP) were produced and mechanically assessed. Furthermore, a novel method of measuring shrinkage in 3D printed components was also developed and used to assess the shrinkage of filaments. The most successful filaments contained 30 wt% harakeke fibers. They had a tensile strength and young's modulus of 39 MPa and 2.8 GPa, respectively, improving tensile strength and young's modulus compared to plain PP filament of 74% and 214%, respectively. However, these properties were seen to reduce on Printing, although fiber reinforcement was still observed to benefit strength and shrinkage. The cause of the mechanical property reduction was assumed to be stress relaxation of the polymer during Printing which is conducted at lower pressure compared to filament production. 30wt% harakeke filaments also underwent the least shrinkage of 0.34%, corresponding to a net reduction of 84% relative to plain.

Zander et al. (2019b) described the reinforcement of recycled polypropylene using cellulose waste materials to generate a green composite feedstock for extrusion-based polymer additive manufacturing. Recycled polypropylene/waste paper, cardboard, and wood flour composites were made using a solid-state shear pulverization. Fourier transform infrared and thermogravimetric analysis were utilized to evaluate the amount of filler incorporated into the 3D-printed materials. Recycled polymer composites had increased levels of filler incorporated in the printed parts compared to the virgin polymer composites based on the thermal gravimetric analysis. The dynamic mechanical analysis showed a ca. 20–30% increase in storage modulus with cellulose materials. Tensile strength was not significantly increased with 10wt % cellulose, but the elastic modulus increased 38% in virgin polypropylene. The analysis of fracture surfaces revealed that failure initiates at the interface, suggesting that the interfacial strength is weaker than the filler strength.

Morales et al. (2021) studied the development of sustainable 3D Printing filaments based on rice husk (RH), an agricultural residue, and recycled polypropylene (rPP) and the influence of fiber weight ratio on the physical, thermal, mechanical, and morphological properties of 3D Printing parts. Thermogravimetric analysis revealed that the composite's degradation process started earlier than the neat rPP due to the lignocellulosic fiber components. Mechanical tests showed that tensile strength increased when using a raster angle of 0° than specimens printed at 90° due to the weaker inter-layer bonding compared to in-layer. Furthermore, interlayer bonding tensile strength was similar for all tested materials. Scanning electron microscope (SEM) images revealed the limited interaction between the untreated fiber and matrix, which led to reduced tensile properties.

Other Composites of Polypropylene Filaments for 3D Printing

Incorporating other reinforcing fibers or fillers with or without treatment into the residual polymer for re-use has improved the strength of the composite, potentially opening needed markets for the material and mitigating the environmental impact (Patty, Cicala and Arcieno, 2021). The global ecological proposal for preserving our planet's natural resources has been gaining attention as an incentive to develop new classes of materials (Kaewunruen and Liao, 2020). In this scenario, composites reinforced with fiberglass and other fillers have also been detached (Rajak et al., 2019).

The use of these composites has also been a trend in additive manufacturing (AM) (Ngo et al., 2018). The growing popularity of AM technology has coincided with increasing environmental awareness, where there is a call for innovative and sustainable composites (Milosevic, Stoof and Pickering, 2017).

Thermoplastic polymers are the most used materials for the AM process due to their relatively low cost and their low melting temperatures (Tan, Zhu and Zhou, 2020). The most common thermoplastic polymers include acrylonitrile butadiene styrene (ABS), polylactic acid (PLA), polycarbonate (PC), polyether ether ketone (PEEK), and nylon (Masood, 2014). However, the resulting pure polymer products can present a lack in terms of strength to produce fully functioning engineering parts (Dickson, Abourayana and Dowling, 2020). Thus, some methodologies like blending, compositing, and branching are commonly used to modify crystallization and improve the behavior of pure polymers. The addition of reinforcing fibers forming composite polymeric materials is a well-known method to improve mechanical properties and aggregate to virgin or recycled polymers (Immonen et al., 2021; Mohammady and HKhairou, 2021; Shulga et al., 2020; Stoof and Pickering, 2017). Polyacrylonitrile (PAN), polyethylene terephthalate (PET), and polypropylene (PP) offer some advantages in fiber-reinforced composite materials and, consequently, have been used for specific applications where their desired properties are applicable (Rajak et al., 2019). The underlying mechanism is that the modifications contribute to the acceleration of crystallization behavior, the decrease of crystal size, and the enhancement of melt strength, thus promoting cell nucleation and stabilizing cell growth (Zhao et al., 2020). Reinforcing materials, such as fibers, have been added into the polymer matrix during Printing to produce a composite structure that typically exhibits improved mechanical properties (Dickson, Abourayana and Dowling, 2020). As PP is frequently used as a structural material in the automotive industry, which requires sufficient stiffness and dimensional stability, these properties can be improved by adding reinforcing fibers (Várdai et al., 2020). Compositing with micro or nanofibers (natural or synthetics) reveals an efficient method to modify crystallization behavior, resulting in improved performance on promoting the foaming ability of PP (Zhao et al., 2020). Table 8 summarizes some recent works published in the literature concerning the preparation of PP-based composites for 3D Printing.

Table 8. PP-based composites for 3D Printing

Reinforcement material	AM methodology	Application	Reference
Glass fiber	FDM and CM	Plastic-based materials	Sodeifian, Ghaseminejad and Yousefi (2019)
Carbon black	FDM	Microwave absorbing materials	Lei et al. (2020)
SiO_2	Vulcanization	-	Mohammady and Khairou (2021)
TiO_2	FDM	-	Kumar, Srivasti and Gautam (2021)
Calcium sulphoaluminate cement	3D printer	Building	Chen et al. (2021)
Graphene	FDM	Eletronics	Shmueli et al. (2020)
Synthetic nanoclay	3D scanning and fusion	-	Chikkangoudar,. Sachidananda and Pattar (2021)
Glass fiber	FDM	Automotive	Joo et al. (2020)
Glass fiber	FFF	Decorative parts	Shulga et al. (2020)
Graphene	FFF	Prototypes	Aumnate et al. (2021)

Sodeifian, Ghaseminejad and Yousefi (2019) performed a comparative study on mechanical, morphological, rheological, and crystallinity properties of pure polypropylene (PP)/glass fiber (GF) and PP/GF composites containing maleic anhydride polyolefin (POE-*g*-MA). Three weight percentages (10, 20, and 30wt%, related to the composites) of POE-*g*-MA were evaluated in their study. The test specimens were provided using 3D Printing (FDM) and compression molding (CM) methods. The authors observed that the presence of POE-*g*-MA enhanced filament flexibility. Tensile tests revealed that the addition of GF increased the modulus and strength of the composite due to its lower flexibility. Furthermore, the composite exhibited decreased modulus and strength and increased flexibility upon the addition of POE-*g*-MA. The specimens prepared via CM exhibited higher strength and modulus values than those prepared via 3D Printing. Rheological studies showed that the presence of POE-*g*-MA resulted in increased storage modulus, reduced viscosity due to the lower value of tanδ. Finally, X-ray diffraction (XRD) analysis indicated higher crystallinity of the specimens prepared via the CM method than those prepared via 3D Printing.

Lei et al. (2020) described carbon black (CB)/polypropylene composites via FDM modeling. The CB content was varied in their study, and the authors observed that when CB content is 10%, uniform dispersion of particles in the matrix was obtained, with a small number of agglomerates. The corresponding composite exhibits excellent microwave absorption and broad adequate bandwidth. The maximum reflection loss value reached − 39.15 dB at 15.08 GHz, and the adequate bandwidth is 5.63 GHz (12.37–18 GHz) with a thickness of 2.1 mm. At the same time, when the thickness is 2.8 mm, the maximum reflection value is up to − 62.6 dB at 10.64 GHz with an adequate bandwidth of 4.56 GHz (8.88–13.44 GHz). The composite containing 10 wt% CB has a conductive loss, polarization loss, and suitable impedance matching, resulting in superior microwave absorption. The authors could conclude that the prepared composites presented improved properties for additive manufacturing applications as 3D printable materials.

Mohammady and Khairou (2021) investigated the effect of silica nanoparticles on the complex shear moduli and molecular dynamics of nanocomposites formed by ethylene propylene diene monomer (EPDM) and polypropylene (PP) blends. The complex shear moduli of the EPDM/PP/SiO_2 nanocomposites and their corresponding neat filler blends were measured at different shear rates in the temperature range from − 95 to 50°C. The effect of filler on the stress-strain and thermal behaviors was also evaluated. The nanocomposites exhibited higher shear moduli than the corresponding pure polymer blends. These results can be attributed to the hindered glass-relaxation dynamics of the EPDM polymer in the nanocomposites.

Kumar, Srivasti and Gautam (2021) described the fabrication of polypropylene -TiO_2 (PP- TiO_2)-based composites via an injection molding process. TiO_2 added varied from 5 to 20 wt% for PP, and the mechanical and thermal characterization of injection-molded PP-TiO_2 composites were evaluated. The results showed that hardness and compressive strength increased as the amount of TiO_2 in the composites increased. Maximum hardness and compressive strength were found in a composite having 20 wt% of TiO_2. The highest flexural strength was achieved13.7% higher than virgin PP in 5 wt% filler loading. No improvement was observed in impact strength due to the hardness of inorganic particles. Composite with 10 wt% presented higher thermal stability and its onset of degradation temperature increased by 80°C. The authors concluded that PP-TiO_2 composites showed important properties for future application in the preparation of 3D specimens.

Chen et al. (2021) prepared 3D printed composites based on polypropylene (PP) and polyvinyl alcohol (PVA)-based fibers reinforced by calcium sulphoaluminate cement composites (CSACCs). The results showed that the PP and PVA fibers enlarge the linear viscoelastic region (LVR) and increase the elastic modulus of CSACCs. The fibers also increased the yield stress and improved the thixotropy of CSACCs. The rheological evaluation showed the structure deformation by the radar map, which indicated that the rheological parameters controlled the printed structure deformation. The authors could conclude that the developed 3D printed fiber-reinforced CSACCs with the controllable rheological parameters can improve the printed structures and toughness for potential application in buildings.

Shmueli et al. (2020) used different techniques to characterize the structure-property relationships of FDM printed isotactic polypropylene/graphene nanocomposites. Rheological measurements of the viscosity indicated that the presence of graphene induced shear thinning during extrusion. With a decrease in the storage modulus, the lamellar structure was observed when 5% graphene was added, which disappeared with 10% graphene. The printed material showed increased thermal conductivity and strong orientational effects for samples with 10% graphene.

Chikkangoudar, Sachidananda and Pattar (2021) described the preparation of PP-nanoclay composites via 3D Printing. The authors observed that the structure of the filaments varies with the addition of nanoclay. 3D Printing filaments were produced for different ratios of polypropylene and nanoclay in the extruder. The nanoclay addition enhanced the dimensional flexibility of polypropylene. However, fragility in filaments and 3D models increased with the increased nanoclay ratio.

Joo et al. (2020) reported the preparation of glass fiber-reinforced PP composite. Pull-out tests were performed according to over-molding conditions of the material to investigate the bond-property changes as a function of process parameter variations. The results showed a change in the behavior of the composite interface due to the cohesive zone model (CZM). These results were applied to a structural simulation, which accurately predicted the mechanical behavior of a prototype front bumper assembles component made of the composites.

Shulga et al. (2020) used the FFF 3D printing technique to manufacture durable, lightweight, and customizable specimens. The composites demonstrated enhanced. The authors used polypropylene filament reinforced by short glass fibers to demonstrate fiber orientation on mechanical tensile properties of the 3D printed specimens. The influence of the printed layer thickness and raster angle on final fiber orientations was investigated using X-ray micro-computed tomography.

The results showed that the better tensile strength of FFF manufactured short fiber-reinforced parts could be obtained along the printing direction, with a layer thickness smaller than the mean fiber length.

Aumnate et al. (2021) described the FFF 3D Printing technique for manufacturing parts in a fast and efficient way. Commercially available PP filaments produce shrinkage and warpage, resulting in instabilities or have poor mechanical properties of the specimens. FFF can therefore only produce prototypes. In their work, the authors solve this instability problem by reinforcing PP with microcapsules.

These microcapsules comprise polylactic acid (PLA) and reduced graphene oxide. The rheological results showed that PP nanocomposites exhibit high melt elasticity, better support printing constructs, and low viscosity, promoting printability, avoiding nozzle clogging, and filament buckling.

Conclusion

This chapter focuses on the re-use of plastic material, linked to shared responsibility for the life cycle of products and the circular economy, to minimize the volume of replaced solid waste and reject and reduce impacts caused to environmental quality. Thus, we report the technical-scientific feasibility of post-consumer polypropylene recycling to obtain 3D printer filaments to define the characteristics necessary for the materials used in FDM technology to guarantee final parts with good properties. We also report using sustainable polypropylene composite filaments and polypropylene composites reinforced with other types of fillers. Thus, it was observed that the recycling process of polymeric materials for 3D Printing is based on several activities such as selective separation of materials, decontamination, purification, milling, re-melting, and extrusion.

The most significant difficulties in recycling homogeneous PP arise from the easy degradability of this polymer during processing and recycling operations. Polypropylene is eventually affected by thermal degradation; the bonds between hydrogen and carbon weaken. Furthermore, mechanical stress and ultraviolet radiation also strongly alter the structure and morphology of PP. Elongation at break and impact strength are the properties most influenced by degradation. Therefore, many researchers have studied ways to improve PP processability by adding loads that can improve mechanical properties, add value to recycled polymer and potentially open needed markets for the material, and mitigate the environmental impact.

REFERENCES

Achilias, D.S. (2012). Material Recycling—Trends and Perspectives. *Intech*, 406.

Ahmed, W., Alnajjar, F., Zaneldin, E., Al-Marzouqi, A.H., Gochoo, M., and Khalid, S. (2020). Implementing FDM 3D Printing Strategies Using Natural Fibers to Produce Biomass Composite. *Materials*, 4065. https://doi.org/10.3390/ma13184065

Akbaş, O.E., Hıra, O., Hervan, S.Z., Samankan, S., and Altınkaynak, A. (2020). Dimensional accuracy of FDM-printed polymer parts. *Rapid Prototyping Journal,* 288–298.

Aliheidari, N., Tripuraneni, R., Hohimer, C., Christ, J., Ameli, A., and Nadimpalli, S. (2017). The impact of nozzle and bed temperatures on the fracture resistance of FDM printed materials. *Behavior and Mechanics of Multifunctional Materials and Composites,* 1016512. https://doi.org/10.1117/12.2260105.

Al-Otaibi. M.S., Alothman, O.Y, Alrashed, M.M., Anis, A., Naveen, J. and Jawaid, M. (2020). Characterization of date palm fiber-reinforced different polypropylene Matrices. *Polymers,* 597.

Alsoufi, M.S., and Elsayed, A.E. (2017). Warping deformation of desktop 3D printed parts manufactured by open source fused deposition modeling (FDM) system. *International Journal of Mechanical and Mechatronics Engineering,* 7–16.

Andena, L., Caimmi, F., Leonardi, L., Nacucchi, M. and De Pascalis, F. (2018). Compression of polystyrene and polypropylene foams for energy absorption applications: a combined mechanical and microstructural study. *Journal of Cellular Plastics*, 49-72.

Aumnate, C., Potiyaraj, P., Saengow, C., Giacomim, A.J. (2021). Reinforcing polypropylene with graphene-polylactic acid microcapsulesfor fused-filament fabrication. *Materials and Design,* 195, 109329, https://doi.org/10.1016/j.matdes.2020.109329.

Baillie, C., Matovic, D., Thamae, T., & Vaja, S. (2011). Waste-based composites - Poverty reducing solutions to environmental problems. *Resources, Conservation and Recycling*, 55, 973 - 978.

Brems, A., Baeyens, R., Dewil R. (2012). Recycling and recovery of post-consumer plastic solid waste in a European context. *Thermal Science*, 16, 669-685.

Camburn, B., and Wood, K. (2018). Principles of maker and DIY fabrication: Enabling design prototypes at low cost. *Design Studies,* 63–88.

Cardoso, R.M., Rocha, D.P., Rocha, R.G., Stefano, J.S., Silva, R.A.B., Richter, E.M., and Muñoz, R.A.A. (2020). 3D-printing pen versus desktop 3D-printers: Fabrication of carbon black/polylactic acid electrodes for single-drop detection of 2,4,6-trinitrotoluene. *Analytica Chimica Acta,* 10–19.

Carneiro, O.S., Silva, A.F. and Gomes, R. (2015) Fused deposition modeling with polypropylene. *Materials & Design*, 768-776.

Chen, H., Zhu, W., Tang, H., and Yan, W. (2021). Oriented structure of short fiber reinforced polymer composites processed by selective laser sintering: The role of powder-spreading process. *International Journal of Machine Tools and Manufacture*, 103703. https://doi.org/10.1016/j.ijmachtools.2021.103703.

Chen, M., Yang, L., Zheng, Y., Li, L., Wang, S., Huang, Y., Zhao, P., Lu, L., Cheng, X. (2021). Rheological behaviors and structure build-up of 3D printed polypropylene and polyvinyl alcohol fiber-reinforced calcium sulphoaluminate cement composites. *Journal of Materials Research and Technology,* 10, 1402-1414, https://doi.org/10.1016/j.jmrt.2020.12.115.

Chethanbabu, H.M., Ramachandra, M. (2021). Evaluation of mechanical properties of polypropylene fibre reinforced epoxy composite filled with silicon carbide particulates. *Materials Today: Proceedings,* https://doi.org/10.1016/j.matpr.2020.09.669. In press.

Chikkangoudar, R.N., Sachidananda, T.G., Pattar, N. (2021). Influence of 3D printing parameters on the dimensional stability of polypropylene/clay printed parts using laser scanning technique, *Materials Today: Proceedings,* 44, 4118-4123, https://doi.org/10.1016/j.matpr.2020.10.456.

Choi, Y.-H., Kim, C.-M., Jeong, H.-S., and Youn, J.-H. (2016). Influence of Bed Temperature on Heat Shrinkage Shape Error in FDM Additive Manufacturing of the ABS-Engineering Plastic. *World Journal of Engineering and Technology*, 186–192.

Chung, T.C.M., (2019) Expanding polyethylene and polypropylene applications to highEnergy areas by applying polyolefin-bonded antioxidants. *Macromolecules*, 5618-5637.

Costa, I.L. M., Pereira, P.H.F., Claro, A.M., Amaral, N.C., Barud, H.S., Ribeiro, R.B., and Mulinari, D.R. (2021). 3D-printing pen from valorization of pine cone residues as reinforcement in acrylonitrile butadiene styrene (ABS): Microstructure and thermal properties. *Journal of Thermoplastic Composite Materials,* https://doi.org/10.1177/08927057211012735.

Cruz Sanchez, F.A., Boudaoud, H., Camargo, M., and Pearce, J.M. (2020). Plastic recycling in additive manufacturing: A systematic literature review and opportunities for the circular economy. *Journal of Cleaner Production,* 121602. https://doi.org/10.1016/j.jclepro.2020.121602.

Dahlbo, H., Poliakova, V., Mylläri, V., Sahimaa, O., and Anderson, R. (2018). Recycling potential of post-consumer plastic packaging waste in Finland. *Waste Management*, 71, 52–61.

De Araujo, F.P., Silva Batista, I.T., De Oliveira, F.C., De Almeida, L.R., De Castro Brito, G., Da Silva Barud, H., Dittz, D., Silva-Filho, E.C., Osajima, J.A., and Lobo, A.O. (2020). Printing composite nanofilaments for use in a simple and low-cost 3D pen. *Journal of Materials Research,* 1154–1162.

Dickson, A.N., Abourayana, H.M., Dowling, D.P. (2020). 3D Printing of Fibre-Reinforced Thermoplastic Composites Using Fused Filament Fabrication—A Review. *Polymers,* 12, 2188, https://doi.org/10.3390/polym12102188.

Exconde, K.J.E., Co, J.A.C., Manapat, J.Z. and Magdaluyo Jr., E.R. (2019). Materials selection of 3D printing filament and utilization of recycled polyethylene terephthalate (PET) in a redesigned breadboard. *Procedia*, 28-32.

Feng, H., Lu, X., Wang, W., Kang, N.G. and Mays, J.W. (2017). Block copolymers: synthesis, self-assembly and applications. *Polymers*, 494.

Ford, S., and Despeisse, M. (2016). Additive manufacturing and sustainability: an exploratory study of the advantages and challenges. *Journal of Cleaner Production*, 1573-1587.

Francis, R. (2016). *Recycling of Polymers: Methods, Characterization and Applications.*

Goodship V. (2007) Plastic Recycling. *Science Progress*. 245-268.

Graziano, A., Jaffer, S. and Sain, M. (2018). Review on modification strategies of polyethylene/polypropylene immiscible thermoplastic polymer blends for enhancing their mechanical behavior. *Journal of Elastomers & Plastics,* 291-336.

Grigore, M. (2017) Methods of Recycling, Properties and Applications of Recycled Thermoplastic Polymers. *Recycling.*

Guo, R., Ren, Z., Bi, H., Song, Y., and Xu, M. Effect of toughening agents on the properties of poplar wood flour/poly(lactic acid) composites fabricated with fused deposition modeling. *European Polymer Journal*, 34-45.

Gupta, S., and Bit, A. (2018). *Rapid prototyping for polymeric gels, Polymeric Gels.* Elsevier Ltd. https://doi.org/10.1016/b978-0-08-102179-8.00016-8

Heidari-Rarani, M., Rafiee-Afarani, M., and Zahedi, A.M. (2019). Mechanical characterization of FDM 3D printing of continuous carbon fiber reinforced PLA composites. *Composites Part B: Engineering,* 107147. https://doi.org/10.1016/j.compositesb.2019.107147.

Immonen, K., Keyriläinen, P.W., Ropponeri, J., Nurmela, A., Metsä-Kortelainen, S., Kaukoniemi, O.V., Kangas, H. (2021). Thermoplastic Cellulose-Based Compound for Additive Manufacturing. *Molecules,* 26, 1701, https://doi.org/10.3390/molecules26061701.

Iunolainen, E. (2017) *Suitability of recycled PP for 3D printing filament.* Degree Thesis, Yrkeshögskolan Arcada. http://urn.fi/URN:NBN:fi:amk-2017112418109

Jafferson, J.M., and Chatterjee, D. (2021). A review on polymeric materials in additive manufacturing. *Materials Today: Proceedings.* https://doi.org/10.1016/j.matpr.2021.02.485.

Joo, S.J., Yu, M.H., Kim, W.S., Lee, J.W., Kim, H.S. (2020). Design and manufacture of automotive composite front bumper assemble component considering interfacial bond characteristics between overmolded chopped glass fiber polypropylene and continuous glass fiber polypropylene composite. *Composite Structures,* 236, 111849, https://doi.org/10.1016/j.compstruct.2019.111849.

Jun, Y.S, Um, J.G., Jiang, G, Lui, G. and Yu A. (2017). Ultra-large sized graphene nano-platelets (GnPs) incorporated polypropylene (PP)/ GnPs composites engineered by melt compounding and its thermal, mechanical, and electrical properties. *Composites Part B*, 218-225.

Kadry, H., Wadnap, S., Xu, C., and Ahsan, F. (2019). Digital light processing (DLP)3D-printing technology and photoreactive polymers in fabrication of modified-release tablets. *European Journal of Pharmaceutical Sciences,* 60–67.

Kaewunruen, S., Liao, P. (2021). Sustainability and recyclability of composite materials for railway turnout systems. *Journal of Cleaner Production*, 285, 124890. https://doi.org/10.1016/j.jclepro.2020.124890.

Karmakar, G. (2020). Regeneration and Recovery of Plastics. *Reference Module in Materials Science and Materials Engineering.*

Kelly, B.E., Bhattacharya, I., Heidari, H., Shusteff, M., Spadaccini, C.M., and Taylor, H.K. (2019). Volumetric additive manufacturing via tomographic reconstruction. *Science,* 1075–1079.

Kiński, W., and Pietkiewicz, P. (2020). Influence of the Printing Nozzle Diameter on Tensile Strength of Produced 3D Models in FDM Technology. *Agricultural Engineering,* 31–38.

Kristiawan, R.B., Imaduddin, F., Ariawan, D., Ubaidillah and Arifin, Z. (2021) A review on the fused deposition modeling (FDM) 3D printing: Filament processing, materials, and printing parameters. *Open Engineering*, 639-649.

Kristiawan, R.B., Imaduddin, F., Ariawan, D., Ubaidillah, and Arifin, Z. (2021). A review on the fused deposition modeling (FDM) 3D printing: Filament processing, materials, and printing parameters. *Open Engineering,* 639–649.

Kumar, D., Srivasti, A., Gautam, V. (2021). Investigation on mechanical and thermo-mechanical properties of injection molded PP-TiO_2 composites. *Materials Today: Proceedings,* 44, 4607-4611.

Lai, Y. and Men, Y. (2018). Polymorph selection during crystallization of random copolymers. *European Polymer Journal*. 218-224.

Lanyi, F.J., Wenzke, N., Kaschta, J., Schubert, D.W. (2019). On the Determination of the Enthalpy of Fusion of α-Crystalline Isotactic Polypropylene Using Differential Scanning Calorimetry, X-Ray Diffraction, and FourierTransform Infrared Spectroscopy: An Old Story Revisited. *Advanced Engineering Materials,* 2019, 1900796.

Lei, L., Yao, Z., Zhou, J., Wei, B., Fan, H. (2020). 3D Printing of carbon black/polypropylene composites with excellent microwave absorption performance. *Composites Science and Technology,* 200, 108479.

Liu, X., Cui, Y. (2021). Multi-scale analysis of the interface structure and failure behaviors for n- SiO2@jute fiber/PP composites. *Composite Structures,* 267, 113865.

Maddah, H.A. (2016) Polypropylene as a promising plastic: A review. *American Journal of Polymer Science,* 1-11.

Majeed, A., Zhang, Y., Ren, S., Lv, J., Peng, T., Waqar, S., and Yin, E. (2021). A big data-driven framework for sustainable and smart additive manufacturing. *Robotics and Computer-Integrated Manufacturing,* 102026. https://doi.org/10.1016/j.rcim.2020.102026.

MarketWatch. (2021). *Fused Deposition Modeling (FDM) 3D Printing Market Research Report 2021.*

Masood, S.H. (2014). Advanced in fused deposition modeling. *Comprehensive Materials Processing,* 10, 69–9, https://doi.org/10.1016/B978-0-08-096532-1.01002-5.

Matthews, C., Moran, F. and Jaiswal, A.K. (2021) A review on European Union's strategy for plastics in a circular economy and its impact on food safety. *Journal of Cleaner Production*, https://doi.org/10.1016/j.jclepro.2020.125263.

Mazzanti, V., Malagutti, L., and Mollica, F. (2019). FDM 3D Printing of Polymers Containing Natural Fillers: A Review of their Mechanical Properties. *Polymers (Basel),* 1094.

McCarthy, E., and Brabazon, D. (2021). Additive Manufacturing for Sustainability of Composite Materials Production. *Reference Module in Materials Science and Materials Engineering.* https://doi.org/10.1016/B978-0-12-819724-0.00067-7.

Mikula, K., Skrzypczak, D., Izydorczyk, G., Warchoł, J., Moustakas, K., Chojnacka, K., and Witek-Krowia, A. (2021). 3D printing filament as a second life of waste plastics—a review. *Environmental Science and Pollution Research*, 12321–12333.

Millet, H., Vangheluwe, P., Block, C., Sevenster, A., Garcia, L. and Antonopoulos, R. (2018) The nature of plastics and their societal usage. *Plastics and the Environment*, 1-20.

Milosevic, M., Stoof, D., Pickering, K.L. (2017). Characterizing the Mechanical Properties of Fused Deposition Modelling Natural Fiber Recycled Polypropylene Composites. *Journal of composites Science,* 1, 7, https://doi.org/10.3390/jcs1010007.

Mirón, V., Ferrándiz, S., Juárez, D., and Mengual, A. (2017). Manufacturing and characterization of 3D printer filament using tailoring materials. *Procedia Manufacturing*, 888–894.

Mohammady, S.Z., Khairou, K.S. (2021). Impact of filler on the soft phase molecular dynamics of PP/EPDM/SiO2 thermoplastic elastomeric nanocomposites. *Egyptian Journal of Petroleum,* In press. https://doi.org/10.1016/j.ejpe.2021.01.004.

Morales, M.A., Martinez, C.L.A., Maranon, A., Hernandez, C., Michaud, V., and Porras, A. (2021). Development and Characterization of Rice Husk and Recycled Polypropylene Composite Filaments for 3D Printing. *Polymers*, 1067.

Nowacki, J., and Sieczkiewicz, N. (2020). Problems of determination of multijet 3d printing distortions using a 3d scanner. *Archives of Materials Science and Engineering*, 30–41.

Patti, A., Cicala, G., Acierno, D. (2021). Eco-Sustainability of the Textile Production: Waste Recovery and Current Recycling in the Composites World. *Polymers,* 13, 134, https://doi.org/10.3390/polym13010134.

Pennafort Jr, L.C.G., Queiroz, W.L.R., Codes, R.N., Deus, E.P. (2014). Caracterização mecânica do compósito PVC/fibra de coco pelos métodos de emissão acústica e de correlação digital de imagem. *Ciência e Tecnologia dos Materiais,* 26, 25-32.

Persad, U., and Mencia, M. (2020). The Value of 3D Printing in Orthopaedics. *Caribbean Medical Journal*, 1–5.

Pickering, K.L. and Stoop, D.B. (2017) Sustainable composite fused deposition modeling filament using post-consumer recycled polypropylene. *Composites Part B: Engineering*, 110-118.

Pinho, A.C., Amaro, A.M. and Piedade, A.P. (2020). 3D Printing goes greener: Study of the properties of post-consumer recycled polymers for the manufacturing of engineering components. *Waste Management*, 426–434.

Ragaert, K., Delva, L., Geem, K. (2017). Mechanical and chemical recycling of solid plastic waste. *Waste Management*, 69, 24–58.

Rajak, D.K., Pagar, D.D., Menezes, P.L., Linul, E. (2019). Fiber-Reinforced Polymer Composites: Manufacturing, Properties, and Applications. *Polymers,* 11, 1667.

Rajamani, D., and Balasubramanian, E. (2019). Investigation of sintering parameters on viscoelastic behaviour of selective heat sintered HDPE parts. *Journal of Applied Science and Engineering*, 39–42.

Rajni, N., Kaul, H., Zhang, L., Rehnberg, B., Stefan, N. (2019). Designing Biobased Recyclable Polymers for Plastics. *Trends in Biotechnology*, 17-18.

Reports and Data. (2021). *Additive Manufacturing Market Analysis by Material Type, by Process, by End-use, and Segment Forecasts to 2027.*

Rigon, D., Ricotta, M. and Meneghetti, G. (2020). A literature survey on structural integrity of 3D printed virgin and recycled ABS and PP compounds. *Procedia Structural Integrity*, 1655–1663.

Salem, A., Lettieri, P., Baeyens, P. (2010) The valorization of plastic solid waste (PSW) by primary to quaternary routes: From re-use to energy and chemicals. *Progress in Energy and Combustion Science*, 36, 103-129.

Sanchez, F.A.C., Boudaoud, H., Hoppe, S., Camargo, M. (2017). Polymer Recycling in an Open-Source Additive Manufacturing Context: Mechanical Issues. *Additive Manufacturing*, 17, 87–105.

Shmueli, Y., Lin, Y.C., Zuo, Z., Guo, Y., Lee, S., Freychet, G., Zhernenkov, M., Kim, T., Tannenbaum, R., Marom, G., Gersappe, D., Rafailovich, M. (2020). In-situ X-ray scattering study of isotactic polypropylene/graphene nanocomposites under shear during fused deposition modeling 3D printing. *Composites Science and Technology,* 196 108227, https://doi.org/10.1016/j.compscitech.2020.108227.

Shubhral, Q.TH. Alam, A.K.M.M. and Quaiyyum, M.A. (2011). Mechanical properties of polypropylene composites: A review. *Journal of Thermoplastic Composite Materials,* 362-391.

Shulga, E., Karamov, R., Sergeichev, I.S., Konev, S.D., Shurygina, L.I., Akhatov, I.S., Shandakov, S.D., Nasibulin, A.G. (2020). Fused Filament Fabricated Polypropylene Composite Reinforced by Aligned Glass Fiber. *Materials,* 13, 3442, https://doi.org/10.3390/ma13163442.

Singh, J., Chawla, K., and Singh, R. (2020). Applications of thermoplastic polymers in 3d Printing. *Reference module in materials science and materials engineering.* //doi.org/10.1016/B978-0-12-820352-1.00010-9.

Singh, J., Chawla, K., and Singh, R. (2020). Applications of Thermoplastic Polymers in 3D Printing. *Reference Module in Materials Science and Materials Engineering, 1*–10.

Singh, N., Hui, D., Singh, R., Ahuja, I., Feo, L., Fraternali, F. (2017). Recycling of plastic solid waste: A state of art review and future applications. *Composites Part B: Engineering*, 115, 409-422.

Singh, N., Hui, D., Singh, R., Ahuja, I., Feo, L., Fraternali, F. (2017). Recycling of plastic solid waste: A state of art review and future applications. *Composites Part B: Engineering*, 115, 409-422.

Sodeifian, G., Ghaseminejad, S., Yousefi, A.K. (2019). Preparation of polypropylene/short glass fiber composite as Fused Deposition Modeling (FDM) filament. *Results in Physics,* 12, 205-222. https://doi.org/10.1016/j.rinp.2018.11.065.

Spoerk, M., Holzer, C. and Gonzalez-Gutierrez, J. (2019). Material extrusion-based additive manufacturing of polypropylene: A review on how to improve dimensional inaccuracy and warpage. *Journal of Applied Polymer Science*, 1-16.

Stansbury, J.W., and Idacavage, M.J. (2016). 3D Printing with polymers: Challenges among expanding options and opportunities. *Dental Materials,* 54–64.

Stegmann, P., Londo, P., and, Junginger, M. (2020). The circular bioeconomy: Its elements and role in European bioeconomy clusters. *Resources, Conservation & Recycling:* X, 100029., 100029.

Stoof, D., and Pichering, K. (2018). Sustainable composite fused deposition modeling filament using recycled pre-consumer polypropylene. *Composites Part B: Engineering,* 110–118.

Tan, L.J., Zhu, W., Zhou, K. (2020) Recent Progress on Polymer Materials for Additive Manufacturing. *Advanced Functional Polymers,* 20, 2003062, https://doi.org/10.1002/adfm.202003062.

Tofail, S.A.M., Koumoulos, E.P., Bandyopadhyay, A., Bose, S., O'Donoghue, L., Charitidis, C. (2018). Additive manufacturing: scientific and technological challenges, market uptake and opportunities. *Materials Today,* 22–37.

Tsioptsias, C., Leontiadis, K., Tzimpilis, E. and Tsivintzelis, I. (2020). Polypropylene nanocomposite fibers: A review of current trends and new developments. *Journal of Plastic Film and Sheeting*, 1-29.

Tzounis, P.N., Argyropoulou, D.V., Anogiannakis, S.D. and Theodorou, D.N., (2018). Tacticity effect on the conformational properties of polypropylene and poly(ethylene–propylene) copolymers. *Macromolecules*. 6878–689.

Várdai, R., Lummerstorferc, V.T., Pretschuhd, C., Jerabekc, M., Gahleitnerc, M., Faludiab, G., Móczóab, J., Pukánszky, B. (2020). Comparative study of fiber reinforced PP composites: Effect of fiber type, coupling and failure mechanisms. *Composites Part A: Applied Science and Manufacturing,* 133, 105895. https://doi.org/10.1016/j.compositesa.2020.105895.

Vardhan, H., Kumar, R. and Chohan, J.S. (2019). Investigation of tensile properties of sprayed aluminum-based PLA composites fabricated by FDM technology. *Materials Today: Proceedings,* 1599–1604.

Vidakis, N., Mountakis, N., Papageorgiou, D., Liebscher, M. and Mechtcherine, V. (2021) Sustainable additive manufacturing: mechanical response of polypropylene over multiple recycling processes. *Journal of Composites Science*. https://doi.org/10.3390/su13010159.

Vidakis, N., Petousis, M., Tzounis, L., Maniadi, A., Velidakis, E., Mountakis, N., Papageorgiou, D., Liebscher, M., and Mechtcherine, V. (2021). Sustainable additive manufacturing: mechanical response of polypropylene over multiple recycling processes. *Sustainability*, 159. https://doi.org/10.3390/su13010159.

Wang, L., Palmer, J., Tajvidi, M., Gardner, D.J., and Han, Y. (2019). Thermal properties of spray-dried cellulose nanofibril-reinforced polypropylene composites from extrusion-based additive manufacturing. *Journal of Thermal Analysis and Calorimetry,* 1069–1077.

Wang, X., Pan, Y, Oin, Y., Voigt, M., Liu, X., Zheng, G., Chen, Q., Schubert, D.W., Liu, C., Shen, C. (2018). Creep and recovery behavior of injection-molded isotactic polypropylene with controllable skin-core structure. *Polymer Testing*, 478-484.

Wickramasinghe, S., Do, T., and Tran, P. (2020). FDM-Based 3D Printing of polymer and associated composite: A review on mechanical properties, defects and treatments. *Polymers,* 1–42.

Wu, W., Geng, P., Li, G., Zhao, D., Zhang, H., and Zhao, J. (2015). Influence of layer thickness and raster angle on the mechanical properties of 3D-printed PEEK and a comparative mechanical study between PEEK and ABS. *Materials,* 5834–5846.

Wu, W., Ye, W., Wu, Z., Geng, P., Wang, Y., and Zhao, J. (2017). Influence of layer thickness, raster angle, deformation temperature and recovery temperature on the shape-memory effect of 3D-printed polylactic acid samples. *Material.* 1-16.

Yaragatti, N., and Patnaik, A. (2020). A review on additive manufacturing of polymers composites. *Materials Today: Proceedings*, 4150–4157.

Zander, N.E., Gillan, M., Burckhard, Z., and Gardea, F. (2019a). Recycled polypropylene blends as novel 3D printing materials. *Additive Manufacturing,* 122-130.

Zander, N.E., Park, J.H., Boelter, Z.R., and Gillan, M.A. (2019b). Recycled Cellulose Polypropylene Composite Feedstocks for Material Extrusion Additive Manufacturing. *ACS Omega*, 13879–13888.

Zhang, D., Qiu, D., Gibson, M.A., Zheng, Y., Fraser, H.L., StJohn, D.H., Easton, M.A. (2019). Additive manufacturing of ultrafine-grained high-strength titanium alloys. *Nature,* 91–95.

Zhao, J., Qiao, Y., Wang, G., Wang, C., Park, C.B. (2020). Lightweight and tough PP/talc composite foam with bimodal nanoporous structure achieved by microcellular injection molding. *Materials and Design*, 195, 109051.

Ziaee, M., and Crane, N.B. (2019). Binder jetting: A review of process, materials, and methods. *Additive Manufacturing,* 781–801.

In: Polypropylene
Editor: Théodore Marleau

ISBN: 978-1-68507-378-7

Chapter 3

INVESTIGATION OF PRO-OXIDANT AGENTS' EFFECTS ON THE NATURAL AGEING AND BIODEGRADATION OF POLYPROPYLENE BAGS

Amina Khelassi and Naima Belhaneche-Bensemra*
Laboratoire des Sciences et Techniques de l'Environnement,
Ecole Nationale Polytechnique, El-Harrach, Alger, Algeria

ABSTRACT

This chapter discusses the effects of calcium, aluminum and zinc stearates as pro-oxidant agents on the degradation of polypropylene (PP) bags under natural weathering.

Samples were collected periodically and characterized. Tensile properties were evaluated, while the structural modifications were investigated using Fourier transform infrared spectroscopy (FTIR) and melt flow index (MFI). Optical microscopy and scanning electron microscopy (SEM) were carried out. Furthermore, thermal properties

* Corresponding Author's E-mail: naima.belhaneche@g.enp.edu.dz.

were investigated using differential scanning calorimetry (DSC) analysis.

The results showed a decrease in elongation and stress at break to very low values, an increase in MFI and carbonyl index over time due to chains' scissions. Changes in melting temperature and crystallinity were also observed after natural weathering.

A biodegradation study was carried out for aged formulations during a year. The results showed that biodegradation of the formulations containing pro-oxidants occurred.

All the considered properties were affected by the presence of the three pro-oxidant agents used in this study. Furthermore, pro-oxidants can be classified as follows:

Zinc stearate > Calcium stearate>Aluminum stearate.

Keywords: aluminum stearate, calcium stearate, zinc stearate, pro-oxidants, PP, degradation

1. INTRODUCTION

Polypropylene (PP) is a polyolefin widely used across the world for a variety of applications, mainly in packaging, due to its excellent properties, that include high tensile strength, hydrophobic nature, mechanical integrity, transparency, ease of processing, and durability [1–4]. On the other hand, some other properties in particular its high molecular weight, hydrophobic nature, the absence of functional groups, and its resistance to microbial attacks, make it non-biodegradable which turns it into a serious problem once its useful life is over [3]. In this context, it is necessary to develop new methods and technologies to make this polymer biodegradable.

The most commonly used method is to incorporate pro-oxidant agents in its structure in form of an organic complex composed of metallic ions mainly stearates [4].

This process is capable of reducing the hydrophobicity and molar weight of PP, favoring degradation by microorganisms [5–6]. According to the literature, biodegradation can be considered a secondary process where microorganisms use these oxidized products to convert H_2O, CO_2, and CH_4 [7, 8]. However, natural weathering caused by solar radiation, moisture, and wind leads to a slow oxidative degradation due to the formation of free radicals [9].

Since oxo biodegradable polymers are not yet as well known as other polymers such as aliphatic polyesters due to the relatively scarce scientific literature, they need to be studied and validated as environmental alternatives [10, 11].

For this purpose, the objective of this study is to compare the effects of three different pro-oxidant agents (POA), calcium, aluminum and zinc stearates on the degradation of polypropylene under natural weather conditions and to investigate the biodegradation of the aged samples in the soil.

2. EXPERIMENTAL

2.1. Products Used and Formulations Produced

The PP/LDPE/$CaCO_3$/POA (73/20/6/1 by weight) bag samples containing different POA ($CaSt_2$, $AlSt_3$ and $ZnSt_2$) and without POA were produced by SASACE company (Société Algeriènne des Sacs Enduits) from Bousmail in the town of Tipaza, Algeria. Polypropylene (PP 506 P) and low density polyethylene (LDPE HP 2022 series) have been provided by Sabic from the Kingdom of Saudi Arabia. Calcium carbonate was imported from Vietnam by the Algerian company Distripol. Zinc stearate ($ZnSt_2$), aluminium stearate ($AlSt_3$) and calcium stearate ($CaSt_2$) were commercial products imported, respectively, from Reagens, Italy and Xatico Benelux, France.

2.2. Natural Exposure

PP bags were exposed to natural Mediterranean climate in Tipaza for 9 months from January to September 2018. Samples were periodically collected for testing and characterization.

2.3. Samples Characterization

2.3.1. Tensile Test

Measurements of tensile properties were taken on Specimens in rectangular form (20 × 2.5 cm), using a reference testometer DBBMCL-500 kg, according to the BS EN ISO 13934-1:1999 at a crosshead speed of 100 mm/min. All the reported values of tensile properties are averaged by five measurements on each specimen.

2.3.2. Melt Flow Index (MFI) Measurement

MFI was measured according to the ISO 1133–1981 (F), using a plastometer ZWICK/ROELL® MFLOW at 190°C and the load was 2.16 kg.

2.3.3. Attenuated Total Reflectance/FT-Infrared Spectroscopy (ATR-FTIR)

To investigate the chemical changes taking place in the material during natural weathering ,Fourier transform infrared (FTIR) spectra were recorded using a SHIMADZU® FTIR spectrophotometer coupled with an attenuated total reflectance (ATR) accessory, into the range of 4000–400 cm−1 with a resolution of 4 cm^{-1}; Two hundred scans were collected for each measurement over the spectral.

The carbonyl index was calculated as the ratio of the maximum absorbance at 1718 cm^{-1} to the absorbance at 2720 cm^{-1} [12].

2.3.4. Differential Scanning Calorimetry Analysis

DSC was carried out using a NETZSCH® STA 409PC/PG Differential scanning calorimeter with a heating rate of 10°C/min in an inert atmosphere of argon. The temperature range scanned was from 20 to 500°C.

The melting enthalpy (ΔH_m) and the melting peak (T_m) were evaluated. The degree of crystallinity (Xc %) was calculated by dividing the melting enthalpy by the enthalpy of 100% crystalline polypropylene ΔH_{0m}, reported as 209 J/g [13].

2.3.5. Optical Microscope Analysis

An optical microscope from OPTIKA® was used to analyze the samples at a magnification of 100 times using 10/0.40 compound lens and transmitted light.

2.3.6. Scanning Electron Microscope Analysis

The surface morphology changes of the PP material during weathering were characterized with scanning electron microscopy (SEM) technique using a JEOL® JSM-6360 model with a 5 kV voltage.

2.4. Soil Burial Degradation Test

This part of the work will focus on the evolution of biodegradation of bags aged for a year in nature. According to the performance of the pro-oxidants used in natural aging, PP-St_2Zn formulation was chosen; PP was taken for comparative purposes. For each formulation, aged and unaged samples, in the form of small squares of the order of (1cm × 1cm) were buried in agricultural soil from the region of Boumerdes (Algeria) at a depth of 2.54 cm according to the literature [14] for a year in crystallizers containing 1 kg of soil each. Samples are watered periodically to replenish any loss caused by evaporation. After each month, the samples were carefully removed from the soil, washed with distilled water, dried for 24 h at 37°C [15], then characterized by FTIR spectroscopy, variation in mass loss and by optical microscopy.

2.4.1. Evolution of Mass Loss

The evolution of the mass loss of the samples was followed as a function of the burial time using the following formula:

$$x = (M_t - M_0) / M_0 \times 100$$

M_t: mass of the sample taken at time t after washing and drying.
M_0: initial mass of the sample.
The balance used is KERN® brand with an accuracy of 10^{-4}g.

2.4.2. Microbiological Analysis

This analysis consists to follow the evolution of microorganisms during burial by the germ count method.

3. Results and Discussion

3.1. Tensile Properties Analysis

Several authors proved that elongation at break can be a good measure to evaluate the ageing of polymers [16-18]. According to the standard (ASTM D3826 - 98 (Reapproved 2008), bags are considered physically degraded when elongation at break is less than or equals to 5%. [19].

Fig. 1 shows the evolution of elongation at break as a function of natural ageing time. After four months of exposure, a slight decrease was noted in elongation at break ε% for all the pro-oxidized formulations (21.17%, 18.05% and 11.16% respectively for PP-$ZnSt_2$, PP-$CaSt_2$ and PP-$AlSt_3$) About 20 days later, suddenly; a drop of ε% was noticed, until reaching values of less than 5%, after that the samples were too fragile for tensile properties testing.

As for PP without pro-oxidant, a 50% loss in ε% was obtained after 210 days after that, elongation remained constant. Along the degradation process, the number of tie molecules connecting the crystalline lamellae decreased significantly which led to a decrease in elongation at break [20-23].

There is also a difference between the degradation speeds of each pro-oxidized formulation depending on the nature of the pro-oxidant in use.

The degradation time of the bags was compared. The PP-$ZnSt_2$ samples were the most degraded after 135 days with an elongation at break of 1.8% and then after 140 days PP-$AlSt_3$ samples and PP-$CaSt_2$ samples, respectively, with elongations at break of 2.5% and 3.5%. Finally, the PP sample without POA which took much longer to degrade.

The evolution of the stress at break with time (Figure 2) confirmed the results obtained with elongation at break, where a drop in the stress at break was observed, as for the elongation at break with different speeds depending on the pro-oxidant in use.

It is to be noted that sometimes the exposed bags contain significant dust content. This can play a protective role by reducing the intensity of photochemical process by absorbing and diffusing harmful solar radiation [24].

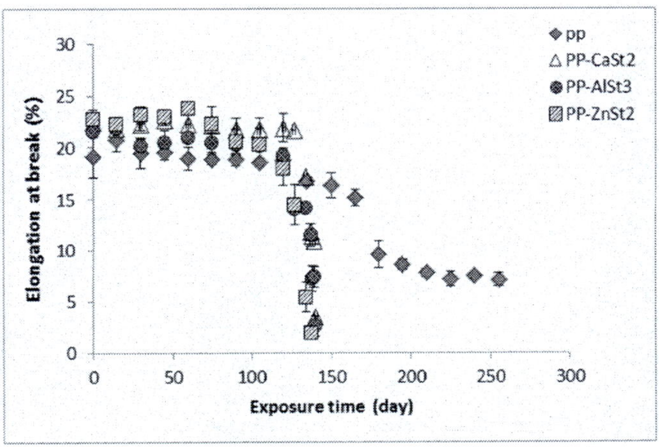

Figure 1. Variation of elongation at break with natural exposure.

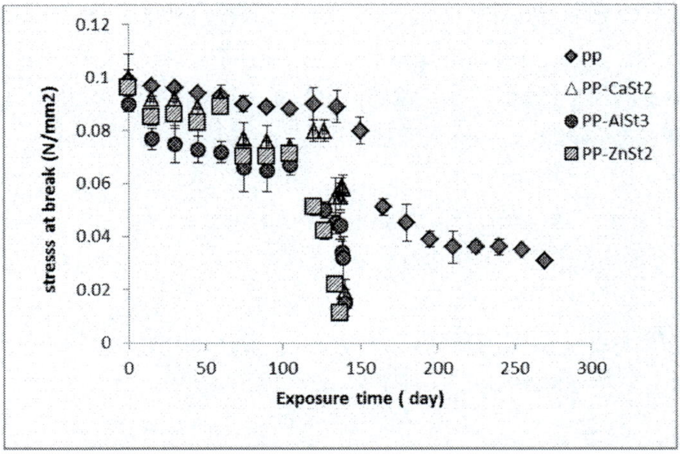

Figure 2. Variation of stress at break with natural exposure.

3.2. MFI Evolution

It is known that MFI increases directly with fluidity and inversely with viscosity. An increased value of MFI translates to a reduced molecular weight (Mw) [25].

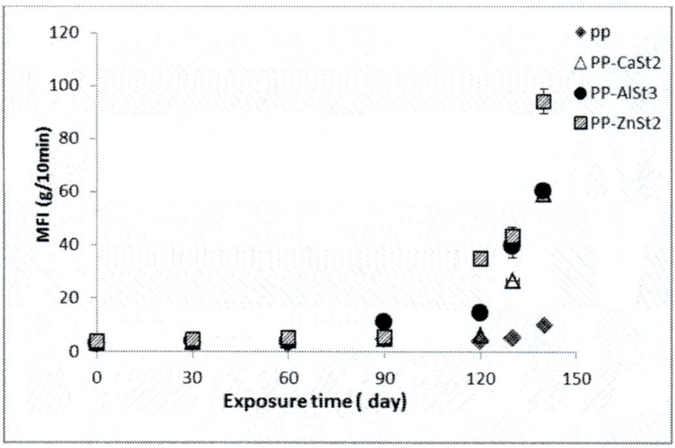

Figure 3. Variation of melt flow index with exposure time.

Figure 3 shows the evolution of MFI with ageing time. MFI remained practically constant for up to 90 days of exposure and then increased. The MFI of pure PP slightly increased (10.30 g/10 min after about 140 days) compared to PP-$ZnSt_2$, PP-$AlSt_3$, PP-$CaSt_2$ which increased up to 94.25, 60.13 and 59.05 g/10 min, respectively, after the same time of exposure. It is thus seen that the longer the aging time, the more MFI increases, which is related to chain scissions reactions caused by the samples' degradation. This results in a drop of molecular masses.

These results are in accordance with those of the tensile properties. The drop of the latter is due to chain scissions. Moreover, the same order of classification of the pro-oxidants agents is noticed.

3.3. FTIR Spectroscopy Analysis

FTIR analysis was used to provide information concerning accumulation of carbonyl containing samples as a function of time. All formulations showed significant changes in the carbonyl zone (1800-1700 cm^{-1}) among which a band centered around 1718 cm^{-1} (Figures 4) [19]. This peak appears after the absorption of different products of degradation like esters, carboxylic acids, ketones, lactones, etc. [26]. The content of carbonyl groups formed with the oxidation of polypropylene was quantified by a carbonyl index obtained by dividing the stretching carbonyl band C=O area at (~1718cm^{-1}) by the area of the reference band. The peak at 2720 $cm^{-1}$was chosen as reference since it is not affected by photo degradation [26].

The results show that there is a significant increase in carbonyl index according to the exposure time (Figure 5) for the three formulations containing pro-oxidant agents which confirms the oxidative degradation reactions which led to the formation of carbonyl products [27-29].

The classic mechanism of photo degradation is initiated by the absorption of light. In the presence of oxygen from air, the radicals formed are oxidized and formed hydro peroxides ROOH, The hydro peroxides formed are very unstable, their decomposition leads to the formation of two kinds of radicals: hydroxyls and unstable alkoxy responsible for chain scission and formation of different carbonyl products.

Degradation of PP containing calcium stearates involves other chemical reactions which accelerate the photo degradation process, due to carbonyl groups present in the $CaSt_2$ structure, according to Fechine et al. [27, 28] who studied the effect of pro oxidant additives on PP photo degradation, these carbonyl groups are generated by photo oxidation which absorbs light and transforms carbonyl groups into biradical forms. These biradical carbonyl groups lead to the formation of radicals by subtracting hydrogen from polypropylene chain.

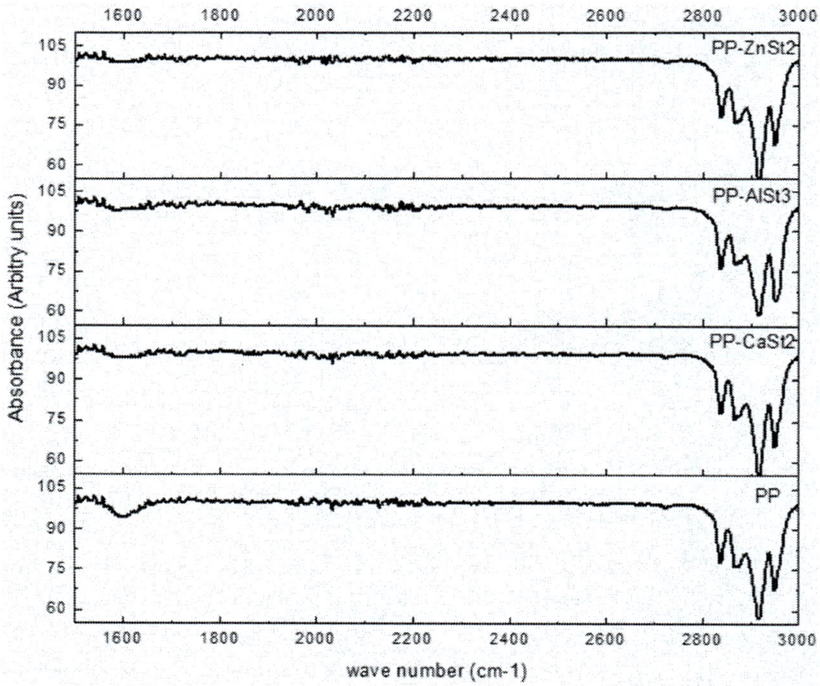

Figure 4. Evolution of FTIR spectra for PP, PP-$CaSt_2$, PP-$AlSt_3$ and PP-$ZnSt_2$ films at 4th month of natural degradation.

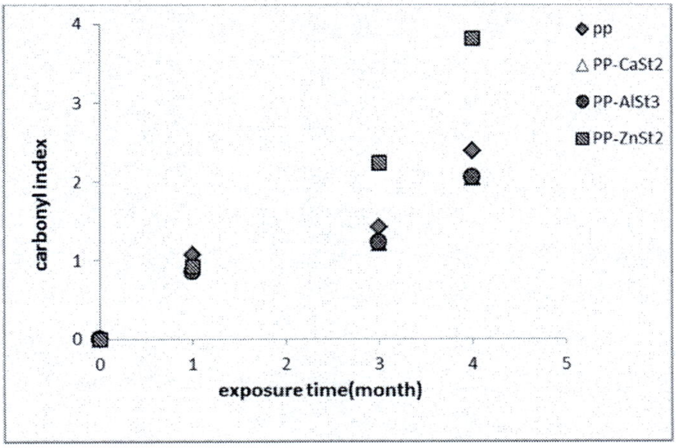

Figure 5. Variation of carbonyl index values with exposure time.

3.4. DSC Characterization

Differential scanning calorimetry was carried out on the PP-$ZnSt_2$ film which showed the most noticeable results during ageing, PP was chosen as a control (Figures 6).

DSC experiments were performed to study the morphological changes in PP samples with and without pro-oxidant agent during natural weathering; The DSC data (heat of fusion (ΔH_m), melting temperature (T_m), and the degree of crystalline (Xc)) of PP and PP-$ZnSt_2$ films are shown in Table 1. PP alone experienced slight changes in melting temperature and crystalline (0.23% lower) after weathering; thus, this material can maintain most of its tensile properties compared to PP-$ZnSt_2$ which showed a decrease in T_m and ΔH_m, which led to a drop in Xc (8.19% lower) [19].; This appreciable reduction in Tm and Xc for weathered PP-$ZnSt_2$ film was already observed by Leong et al. [30] and Beg et al. [31]. On the other hand, an increase in crystallinity for weathered PP samples was reported [32, 33]. It is generally accepted that oxidation, although mainly occurring in the amorphous phase because of its higher oxygen permeability, may also take place in the interface of crystallites by chain scissions [34] allowing the smaller molecules to rearrange, and leading to an increase in the degree of crystallinity [35]. Reactions of chain scissions can reach the tie molecules (chain crossing the amorphous phase from one crystalline lamella to another) if they continue to occur leading to modification of crystallinity [36]. Besides, the increasing of photodegradation products like carbonyl groups can reduce the molecular regularity of polypropylene and limits its secondary crystallisation [37–39]. So, in this work, the more advanced stage of degradation can be the cause of crystallinity reduction.

Table 1. DSC data of PP and PP-$ZnSt_2$ films before and after natural ageing

Time (months)	PP			PP-$ZnSt_2$		
	T_m (°C)	ΔH_m	X_c (%)	T_m (°C)	ΔH_m	X_c (%)
0	158.89	15.68	7.51	157.48	29.63	14.18
1	157.64	14.77	7.07	159.38	16.78	8.30
2	159.33	21.62	10.35	154.00	16.86	8.70
3	156.00	15.20	7.28	155.00	12.50	5.99

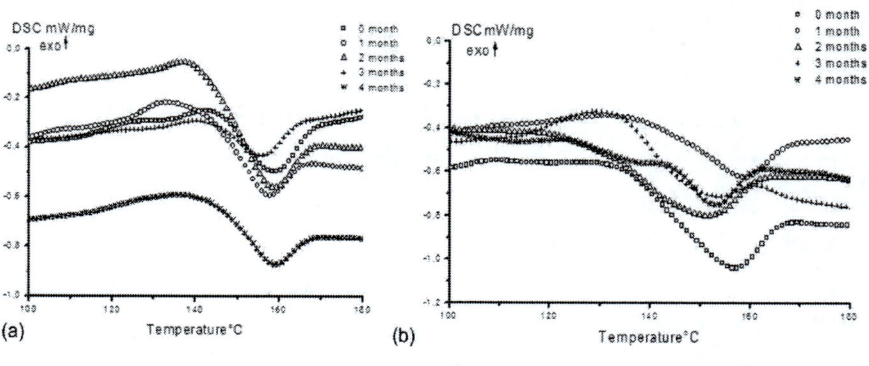

Figure 6. DSC thermograms of PP (a) and PP-$ZnSt_2$ (b) films after natural ageing.

3.5. Optical Microscopy Analysis

Figures 7 illustrate the optical images of PP; PP-$CaSt_2$, PP-$AlSt_3$ and PP-$ZnSt_2$ films at the 4th month of natural degradation.

The appearance of cracks is noticed in the films containing pro-oxidants, these cracks cause a drop of mechanical properties, whereas cracks were not observed in PP without pro-oxidant [19].

On the other hand, a considerable amount of bacterial colonies was formed in the case of samples containing pro-oxidants. The sample without pro-oxidant did not present a significant number of colonies, this confirms a slight degradation of the film without pro-oxydant.

Bonhomme et al. [29] have encountered the same situation and found biofilm on the PE surface.

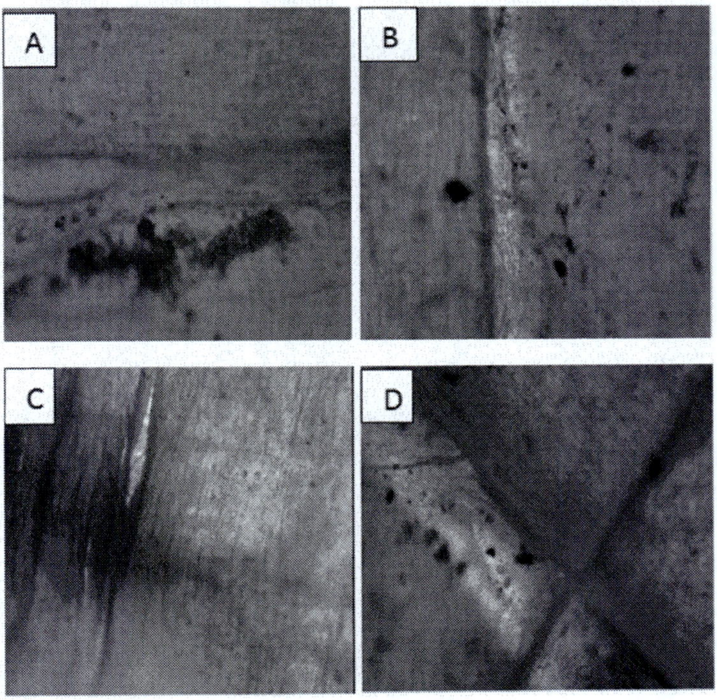

Figure 7. Optical microscopy images of (A): PP, (B): PP-Ca St_2, (C): PP-Al St_3 and (D): PP-Zn St_2 formulations at 4th month of natural degradation.

3.6. Scanning Electron Microscopy Analysis

Scanning electron microscope images were taken for the PP-$ZnSt_2$ which the fastest degradation rate according to the previous results, before and after ageing (Figure 8). The surface morphology of the aged film shows cracks which confirm the degradation of the sample [19].

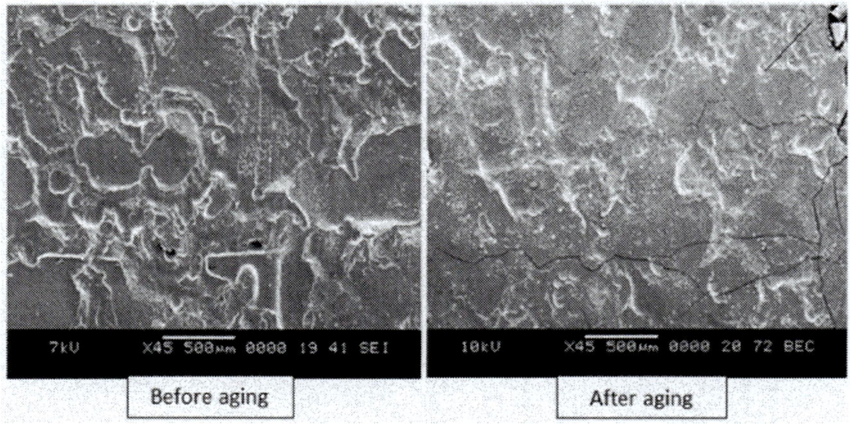

Figure 8. Scanning electron microscopy images of PP-$ZnSt_2$ formulation.

3.7. Soil Burial Degradation

3.7.1. Variation of Mass

Figure 9 illustrates the change in mass loss of PP and PP-$ZnSt_2$ formulations, whether or not they have undergone natural aging and buried in the soil for 12 months.

After 12 months of burial in the ground, the unaged films showed a slight loss of weight (-0.85% and -4.35 for the formulations PP and PP-$ZnSt_2$, respectively).

After the same period of burial, aged PP-$ZnSt_2$ showed a greater mass loss (-56.32%) unlike the PP which showed a loss of weight of -3.40%, which confirms the effect of pro-oxidizing agents on the biodegradation of PP.

This loss is due either to the crumbling of the sample under the action of the soil or to the biodegradation of the material under the action of microorganisms [40].

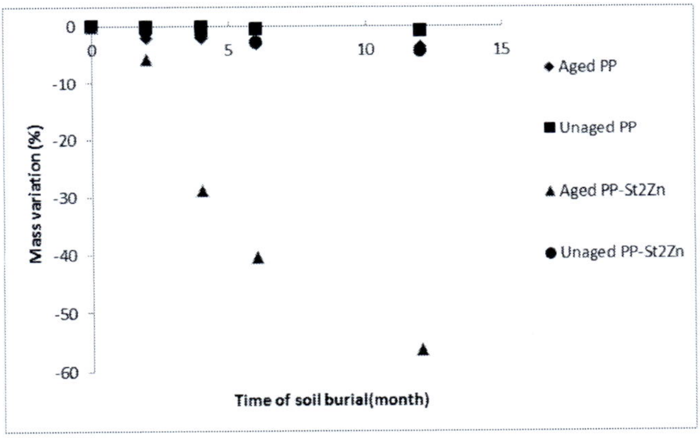

Figure 9. Mass variation as a function of time of soil burial.

3.7.2. Fourier Transform Infrared Spectroscopy Analysis

The structural changes that occurred during the burying of the samples were followed by FTIR spectroscopy.

Figures 10 to 13 show the infrared spectra of the aged and unaged PP and PP-$ZnSt_2$ formulations after 12 months of burial in the soil.

No structural change was recorded in the case of formulations not naturally aged and buried in the soil (Figure 10 and 12).

The aged PP-$ZnSt_2$ formulation (figure 13) showed the appearance of new bands in the 1600-1800 cm-1 zone, which would have been caused by the formation of several carbonyl species resulting from biodegradation under the action of microorganisms present in the soil.

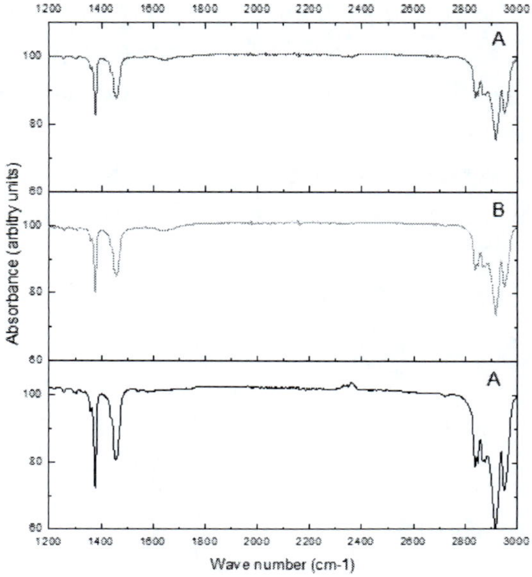

Figure 10. FTIR spectras for PP unaged formulation, A: reference, B: 6 months of burial, C: 12 months of burial in the soil.

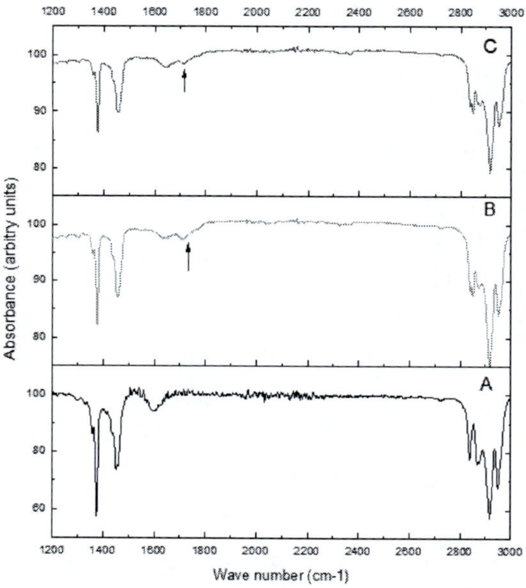

Figure 11. FTIR spectras for PP aged formulation, A: reference, B: 6 months of burial, C: 12 months of burial in the soil.

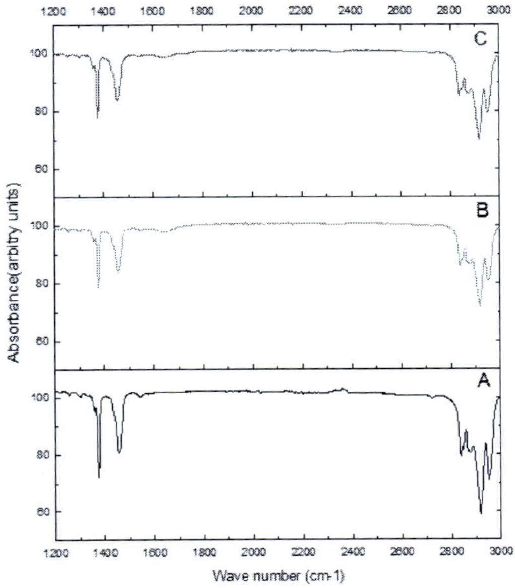

Figure 12. FTIR spectras for PP-$ZnSt_2$ unaged formulation, A: reference, B: 6 months of burial, C: 12 months of burial in the soil.

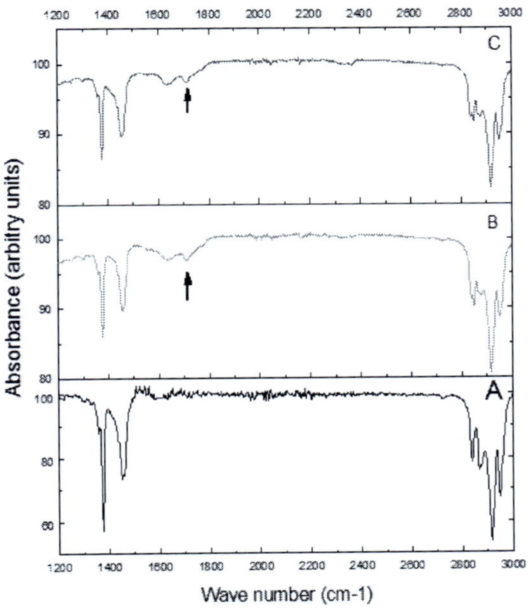

Figure 13. FTIR spectras for PP-$ZnSt_2$ aged formulation, A: reference, B: 6 months of burial, C: 12 months of burial in the soil.

3.7.3. Enumeration of Germs

Table 2 shows the evolution of the number of colonies formed before and after 12 months of burial.

An increase in the number of colonies was observed in the case of the pro-oxidized formulation but it was more noticeable in the case of the aged films, which confirms the existence of a biodegradation phenomenon.

PP alone showed a low number of colonies compared to pro-oxidized film; this may be due to the resistance of this polymer to microorganisms' attacks.

Table 2. Evolution of the number of colonies in CFU / g as a function of burial time

	Before burial	12 months of burial	
		Not aged	Aged
PP	$7,0. 10^2$	$2,7. 10^3$	$4,5. 10^3$
$PP-St_2Zn$	$7,0. 10^2$	$3,7. 10^3$	$9,0. 10^3$

3.7.4. Optical Microscopy Analysis

Figures 14 to 17 show optical images of unaged and aged PP and PP-$ZnSt_2$ formulations, respectively, before and after burial in soil.

We notice in the optical images of the unaged formulations that no morphological change has taken place, except for the appearance of a few bacterial colonies.

For the aged PP-$ZnSt_2$ formulation, microcracks were observed after burying, in addition to the appearance of bacterial colonies, unlike the unaged PP which did not show any morphological changes. This confirms the results found previously [19].

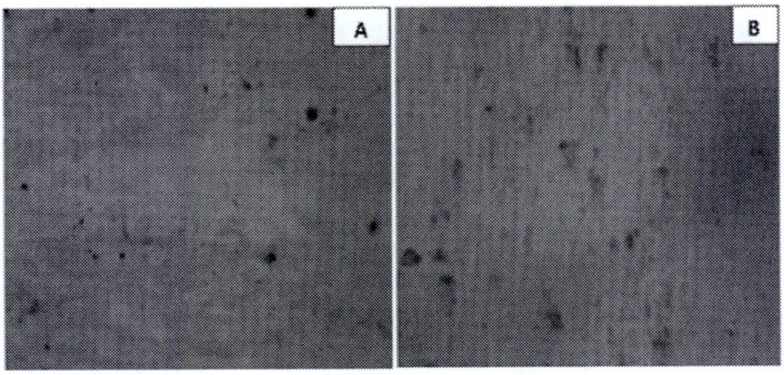

Figure 14. Optical microscopy images of unaged PP, (A): Before soil burial, (B): After soil burial.

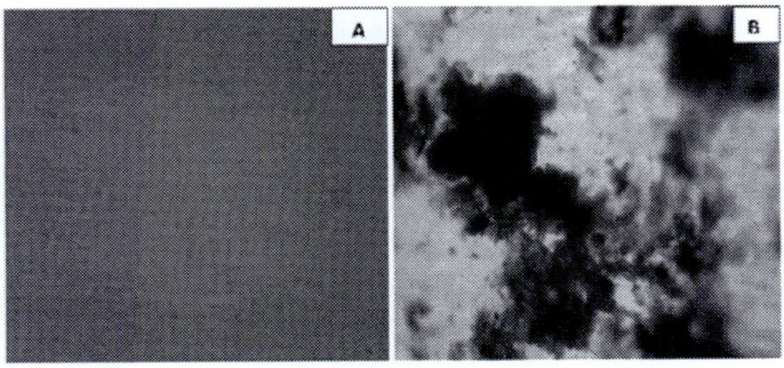

Figure 15. Optical microscopy images of aged PP, (A): Before soil burial, (B): After soil burial.

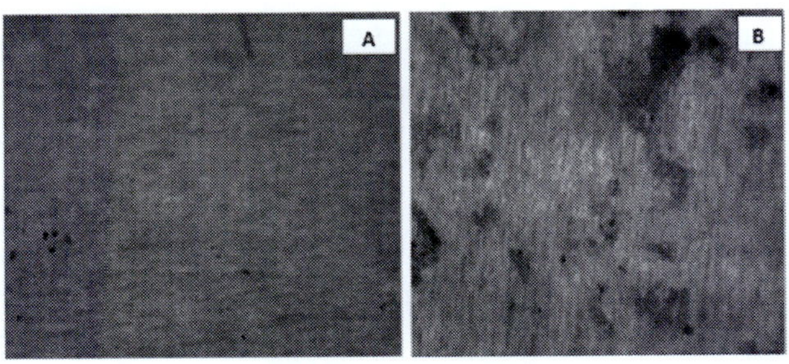

Figure 16. Optical microscopy images of unaged PP-$ZnSt_2$, (A): Before soil burial, (B): After soil burial.

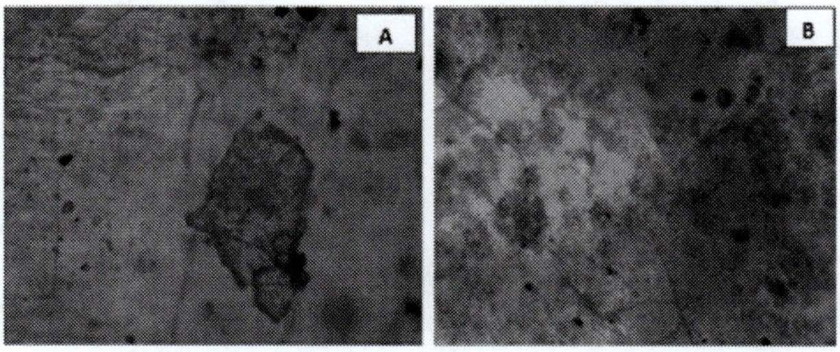

Figure 17. Optical microscopy images of aged PP-$ZnSt_2$, (A): Before soil burial, (B): After soil burial.

CONCLUSION

Natural aging and burial in the soil were carried out in order to study the photo-degradation and biodegradation of pro-oxidized films.

The characterization of films that have undergone natural aging allowed us to draw the following conclusions:

A decrease in tensile properties, a drop in weight and an increase in carbonyl number was caused by chain scission reactions.

Optical microscopy and scanning electron microscopy showed the expansion of cracks on the surface of the films and the appearance of microorganisms' colonies.

A decrease in melting temperature, heat of fusion and crystallinity caused by photo-degradation was also observed in the case of PP-$ZnSt_2$ film.

Finally, the addition of pro-oxidants in the PP made the natural photo-degradation much faster and the effects of pro-oxidants can be classified as follows:

Zinc stearate > Calcium stearate > Aluminum stearate

Biodegradation of the photo-degraded formulations was followed by a soil burial test for one year.

A significant mass loss was noticed for the photo-degraded pro-oxidized formulation.

FTIR spectroscopic analysis showed a slight structural change in the formulation containing a pro-oxidant, indicating its biodegradation.

The enumeration of the colonies of bacteria showed a growth of the latter in the case of the photo-degraded pro-oxidized formulation.

Optical microscopy showed the formation cracks and colonies of bacteria.

Finally, this work made it possible to show that pro-oxidized polypropylene films are oxo-biodegradable.

Acknowledgments

This work was supported by DGRSDT (Direction Générale de la Recherche Scientifique et du Développement Technologique, Algeria).

The authors would like to express their gratitude to Mrs K. Taoussi, Mr K. Bentifour, N. Benberim and R. Draachour from the Algerian company SASACE (Société Algeriènne des Sacs Enduits) for assistance in the preparation of the films.

References

[1] Constantinou. A et al. 2020. "Identification of commercial oxo-biodegradable plastics: study of UV induced degradation in an effort to combat plastic waste accumulation." *Journal of Polymers and the Environment* 28:2364–2376. https://link.springer.com/content/pdf/10.1007/s10924-020-01776-x.pdf.

[2] Hayoune. F et al. 2020. "Thermal decomposition kinetics and lifetime predictionof a PP/PLA blend supplemented with iron stearate during artificial aging." *Thermochim Acta* 690:178700. https://www.sciencedirect.com/science/article/pii/S004060312030294X.

[3] Sunil. S, Sanjeev. A, and Haripada B. 2020. "Preparation and Characterization of Oxo-degradable Polypropylene Composites Containing a Modified Pro-oxidant." *Journal of Polymers and the Environment.* doi: 10.1007/s10924-020-01910-9.

[4] Albertsson, A.C, Barenstedt. C, and Karlsson, S. 1994. "Abiotic degradation products from enhanced environmentally degradable polyethylene." *Acta Polymerica* 45(2), 97–103. doi: 10.1002/actp. 1994.010450207.

[5] Shah. A, et al. 2008. "Biological degradation of plastics: a comprehensive review." *Biotechnology Advances* 26:246–265. https://www.sciencedirect.com/science/article/pii/S0734975008000 141.

[6] Yang. J, et al. 2014. "Evidence of polyethylene biodegradation by bacterial strains from the guts of plastic-eating wax worms." *Environmental Science and Technology* 48:13776–13784. https://pubs.acs.org/doi/abs/10.1021/es504038a.

[7] Ojeda. T, et al. 2011. "Degradability of linear polyolefins under natural weathering." *Polymer Degradation and Stability* 96:703–707. https://www.sciencedirect.com/science/article/pii/S01413910 10004581.

[8] Albertsson. A, Kalrsson. S. 1994. "Environment-adaptable polymers." *Polymer Degradation and Stability* 41:345–349. https:// www.sciencedirect.com/science/article/pii/014139109390082T.

[9] Gewert. B, Plassmann. M, and Macleod. M. 2015. "Pathways for degradation of plastic polymers floating in the marine environment." *Environmental Science-Processes & Impacts* 17:1513–1521.

[10] Chiellini. E, Corti. A, and D'Antone. S. 2007. "Oxo-biodegradable full carbon backbone polymers–biodegradation behaviour of thermally oxidized polyethylene in an aqueous medium." *Polymer Degradation and Stability* 92(7), 1378–1383. doi: 10.1016/j. polymdegradstab.2007.03.007.

[11] Koutny.M, Lemaire. J, and Delort. A. M. 2006. "Biodegradation of polyethylene films with prooxidant additives." *Chemosphere* 64(8), 1243–1252. doi: 10.1016/j.chemosphere.2005.12.060.

[12] Bagheri. R. 2009. "Study of the effect of γ-absorbed dose on degradation of polypropylene/starch-filled polymer containing photo-initiators by spectroscopic methods." *Radiation Physics and Chemistry*, 78(9), 765-769. https://www.sciencedirect.com/science/article/pii/S0969806X09001583.

[13] Rabello, M. S. and White, J. R. 1997. "Crystallization and melting behaviour of photodegraded polypropylene– 1. Chemi-crystallization." *Polymer*, 38(26), 6379-6387. https://www.sciencedirect.com/science/article/pii/S0032386197002139.

[14] Masood. F, Yasin. T, and A. Hameed. 2014. "Comparative oxo-biodegradation study of poly-3 hydroxybutyrate-co-3-hydroxyvalerate/polypropylene blends in controlled environments." *International Biodeterioration and Biodegradation*, 87, 18,. https://www.sciencedirect.com/science/article/pii/S0964830513003703.

[15] Torikai. A, et al. 1990. "Photodegradation of polyethylene: factors affecting photostability." *Journal of Applied Polymer Science*, 40(9-10), 1637-1646. doi: 10.1002/app.1990.070400919.

[16] Torikai. A, et al. 1990. "Photodegradation of polyethylene: factors affecting photostability." *Journal of Applied Polymer Science*, 40(9-10), 1637-1646. doi: 10.1002/app.1990.070400919.

[17] Hoekstra. H, Spoormaker. J, and Breen. J. 1997. "Mechanical and morphological properties of stabilized and non-stabilized HDPE films versus exposure time. Die Angewandte Makromolekulare Chemie." *Applied Macromolecular Chemistry and Physics*, 247(1), 91-110.

[18] Bensaad. F, Belhaneche-Bensemra. N. 2018. "Effects of calcium stearate as pro-oxidant agent on the natural aging of polypropylene." *Journal of Polymer Engineering*, 38 (8), 715-721. doi: 10.1515/polyeng-2017-0391.

[19] Khelassi. A, Belhaneche-Bensamra. N. 2020. "Investigation of pro-oxidant agents' effects on the ageing of polypropylene bags."

International Journal of Environmental Studies, 77, 264-274. doi: 10.1080/00207233.2020.1727231.

[20] Naddeo. C, Guadagno. L, and Vittoria. V. 2004. "Photooxidation of spherilene linear low-density polyethylene films subjected to environmental weathering. 1. Changes in mechanical properties." *Polymer Degradation and Stability*, 85(3), 1009-1013. doi: 10.1016/j.polymdegradstab.2003.04.005.

[21] Ojeda. T, et al. 2011. "Degradability of linear polyolefins under natural weathering." *Polymer Degradation and Stability*, 96(4), 703-707. doi: 10.1016/j.polymdegradstab.2010.12.004.

[22] Guadagno. L, et al. 2001. "Chemical and morphologial modifications of irradiated linear low density polyethylene (LLDPE)." *Polymer Degradation and Stability*, 72(1), 75-186. doi: 10.1016/S0141-3910(01)00024-6.

[23] Suresh. Et al. 2011. "Influence of thermal oxidation on surface and thermo-mechanical properties of polyethylene." *Journal of Polymer Research*, 18(6), 2175-2184. doi: 10.1007/s10965-011-9628-0.

[24] Jan-Chan. H, Junke. X. 2003. "Thermal degradation of polypropylene in a capillary rheometer." *International Journal of Polymeric Materials and Polymeric Biomaterials*, 52, 203-209. doi: 10.1080/00914030304896.

[25] Aslanzadeh. S, Kish. M. H. 2005. "Photodegradation of polypropylene thermal bonded non-woven fabric." *Polymer Degradation and Stability*, 90(3), 461-470. doi: 10.1016/j. polymdegradstab.2005.04.009.

[26] Hakkarainen. M, Albertsson. A.C. 2004. "Environmental degradation of polyethylene." *Advances in Polymer Science*, 169, 177-199.

[27] Fechine. G, Demarquette. N. 2008. "Cracking formation on the surface of extruded photodegraded polypropylene plates." *Polymer Engineering and Science*, 48(2), 365-372. doi: 10.1002/(ISSN) 1548-2634.

[28] Fechine. G, et al. 2009. "Effect of UV radiation and pro-oxidant on PP biodegradability." *Polymer Engineering & Science*, 49(1),123-128. doi: 10.1002/pen.v49:1.

[29] Bonhomme. S, et al. 2003. "Environmental biodegradation of polyethylene." *Polymer Degradation and Stability*, 81(3), 441-452. doi: 10.1016/S0141-3910(03)00129-0.

[30] Leong. Y, et al. 2004. "Characterization of talc/calcium carbonate filled polypropylene hybrid composites weathered in a natural environment." *Polymer Degradation and Stability*, 83(3), 411-422. doi: 10.1016/j.polymdegradstab.2003.08.004.

[31] Beg. M. D. H, Pickering. K. L. 2008. "Accelerated weathering of unbleached and bleached Kraft wood fibre reinforced polypropylene composites." *Polymer Degradation and Stability*, 93(10), 1939-1946. doi: 10.1016/j.polymdegradstab.2008.06.012.

[32] Li. J, et al. 2008. "Natural photo-aging degradation of polypropylene nanocomposites." *Polymer Degradation and Stability*, 93(1), 84-89. doi: 10.1016/j.polymdegradstab.2007.10.022.

[33] Craig. I, White. J, and Kin. P. C. 2005. "Crystallization and chemi-crystallization of recycled photo-degraded polypropylene." *Polymer*, 46(2), 505-512. doi: 10.1016/j.polymer.2004.11.019.

[34] Lacey. D, Dudler. V. 1996. "Chemiluminescence from polypropylene. Part 1: Imaging thermal oxidation of unstabilised film." *Polymer Degradation and Stability*, 51(2), 101-108. doi: 10.1016/0141-3910(95)00151-4.

[35] Yakimets. I, Lai. D, and Guigon. M. 2004. "Effect of photo-oxidation cracks on behaviour of thick polypropylene samples." *Polymer Degradation and Stability*, 86(1), 59-67. doi: 10.1016/j.polymdegradstab.2004.01.013.

[36] Stark. N. M, Matuana. L. M. 2004. "Surface chemistry and mechanical property changes of wood-flour/ high-density-polyethylene composites after accelerated weathering." *Journal of Applied Polymer Science*, 94(6), 2263-2273. doi: 10.1002/(ISSN)1097-4628.

[37] Chan. J, Balke. S. 1997. "The thermal degradation kinetics of polypropylene: Part III. Thermogravimetric analyses." *Polymer Degradation and Stability*, 57(2), 135-149. doi:10.1016/S0141-3910(96)00160-7.

[38] Rabello. M, White. J. 1997. "Crystallization and melting behaviour of photodegraded polypropylene- I. Chemi-crystallization." *Polymer* 38(26), 6379-6387. doi: 10.1016/S0032-3861(97)00213-9.

[39] Tang. L, Wu. Q, and Qu. B. 2005. "The effects of chemical structure and synthesis method on photodegradation of polypropylene." *Journal of Applied Polymer Science*, 95(2), 270-279. doi: 10.1002/(ISSN)1097-4628.

[40] Rabello. M, White. J. 1997. "Crystallization and melting behaviour of photodegraded polypropylene—II. Re-crystallization of degraded molecules." *Polymer*, 38(26), 6389-6399. doi: 10.1016/S0032-3861(97)00214-0.

BIOGRAPHICAL SKETCH

Naima Belhaneche-Bensemra

Affiliation: Ecole Nationale Polytechnique.

Education: PhD degree in Macromolecular Materials and Composites from Claude Bernard University, Lyon I, France, 1989; Engineering degree in Polymeric Materials from Ferhat Abbas University, Sétif, Algeria, 1985.

Research and Professional Experience: Since 1989, Dr Naima Belhaneche-Bensemra is professor at the Environment Engineering Department of the National Polytechnic School of Algiers (Algeria) where she is in charge of various lectures. She is also the head of the "Materials and Environment" team in Laboratory of Science and Environment Techniques since 2000.

Dr Naima Belhaneche- Bensemra is the author or co-author of over one hundred papers in international refereed journals and over one hundred papers in conference proceedings, author of chapters in books and referee of about thirty journals.

Professional Appointments: Professor, Research director, Director of the Laboratory of Science and Environment Techniques in National Polytechnic School, Algiers.

Publications from the Last 3 Years:

1. Properties Investigation of Epoxidized Sunflower Oil as Bioplasticizer for Poly (Lactic Acid), *Journal of Polymers and the Environment*, 2021.
2. Optimisation of Operatory Conditions for Synthesis of Sunflower Oil Biobased Polyols Using Design of Experiments and Spectroscopic Methods, *Journal of Polymers and the Environment*, 2021. doi: 10.1007/s10924-020-01917-2.
3. Influence of Different Diss Fiber Treatments over the Properties of Poly Propylene/Recycled and Regenerated Low Density Polyethylene Based Biocomposites, *Journal of Polymers and the Environment*, 2021. doi: 10.1007/s10924-020-01877-7.
4. Elaboration and Characterization of Polyurethane Foams Based on Renewably Sourced Polyols, *Journal of Polymers and the Environment*, 2020. doi: 10.1007/s10924-020-01833-5.
5. Low-Density Polyethylene/Poly(Lactic Acid) Blends Reinforced by Waste Wood Flour, *Journal of Vinyl and Additive Technology*, 2020 doi: 10.1002/vnl.21759.
6. Investigation of Polyethylene Pipeline Behavior after 30 Years of Use in Gas Distribution Network, *Journal of Materials Engineering and Performance*, 2020. doi: 10.1007/s11665-020-05118-9.
7. Antimony leaching from PET plastic into bottled water in Algerian market, *Environmental Monitoring and Assessment*, 2019 doi: 10.1007/s10661-019-7891-4.

8. Investigation of additives migration from salted serum bags, *Polymer Testing*, 2019. doi: 10.1016/j.polymertesting.2019.106118.
9. Modification of Diss Fibers for Biocomposites Based on *Recycled Low-Density Polyethylene and Polypropylene Blends, Waste and Biomass Valorization*, 2019. doi: 10.1007/s12649-018-0225-x.
10. Preparation of epoxidized sunflower oil metal soap derivatives and their use asheat stabilizers for polyvinyl chloride, *Turkish Journal of Chemistry*, 2019. doi: 10.3906/kim-1806-31.
11. Evaluation of the effects of acetyl tributyl citrate as bio-based plasticizer on the physical, thermal, and dynamical mechanical properties of poly(vinyl chloride)/polymethyl methacrylate blends, *Journal of Vinyl and Additive Technology*, 2019. doi: 10.1002/vnl.21646.

In: Polypropylene
Editor: Théodore Marleau
ISBN: 978-1-68507-378-7

Chapter 4

POLYPROPYLENE BASED COMPOSITES: A BETTER ALTERNATIVE TO CONVENTIONAL RADIATION SHIELDING MATERIALS

Mehmet Kılıç, Yaşar Karabul and Zeynep Güven Özdemir*
Physics Department,
Yıldız Technical University, İstanbul, Turkey

ABSTRACT

Polypropylene (PP) is a thermoplastic polymer with a wide range of uses, from the automotive industry to textile and food packaging. In addition to these known application areas of PP, the use of PP in radiation shielding applications is a virgin area. Although there are currently limited scientific studies on PP-based radiation shielding materials, the commercially advantageous properties of PP such as its low cost, lightweight, easy process, high chemical resistance make PP

* Corresponding Author's E-mail: zgozdemir@gmail.com.

attractive for using it in the production of radiation shielding materials. From this point of view, the success of composites containing polypropylene in radiation shielding has been reviewed within the scope of the current scientific literature in the first part of the chapter. The second part of the chapter has been devoted to prepare the novel PP-based microstructured composites consisting of natural mineral and ceramic reinforcing particles for radiation shielding. For this purpose, Yahyalı Stone (YS) and the bismuth-based copper oxide layered ceramic micro powders have been used for reinforcing particles for the PP matrix. The radiation shielding performances of the bulk PP/YS and PP/$Bi_2Sr_2Ca_2Cu_3O_{10+x}$ composites with different mass percentages of micro reinforcing particles have been obtained experimentally and theoretically. The radiation shielding abilities of the samples have been evaluated in the context of linear attenuation coefficient (μ), half-value layer (HVL), mean free path (MFP), effective atomic number (Z_{eff}), and effective electron density (N_{eff}). The experiments have been carried out by experimental setup equipped with NaI(Tl) detector and the Ba-133, Cs-137, and Co-60 point radioactive sources that emit gamma rays with energies ranging from 81 keV to 1332 keV. The theoretical calculations have also been realized by using WinXCom software. As a result, it has been concluded that the highest reinforcing particle contributed PP composites show significant radiation attenuation against X-rays and low energetic gamma-rays. In this context, the mentioned composites can be considered to have the potential to replace conventional radiation-shielding materials such as lead and concrete, which have serious disadvantages in terms of toxicity and weight.

Keywords: polypropylene, composites, radiation shielding, X-ray, gamma-ray

INTRODUCTION

Polypropylene (PP), one of the thermoplastic polymers, is produced as a result of the polymerization of propylene monomers. PP, an extremely inexpensive polymer, is used in a wide range of industries such as automotive, textile, and packaging. PP, a synthetic polymer, was first synthesized in 1951. Afterward, its resin has been synthesized in 1954. In the few years that followed, it became a very low-cost commercial polymer across Europe. Nowadays, it is the most widely produced polymer in the world.

Besides these very well-known applications of PP, this chapter is particularly focused on researching and developing a feature of PP that may be considered hidden: "*Radiation Shielding.*" For this purpose, in the first sections, the factors that determine the radiation shielding performance of material have been briefly introduced, then the results of some selected researches in radiation shielding performed with PP and PP-based composites have been mentioned. After emphasizing the point that PP-based composites have reached in the context of radiation shielding, the chapter has focused on the experimental and theoretical investigation of the radiation shielding properties of two new PP-based composites, one is reinforced with a natural stone, and the other is a ceramic. In the last stage, the results have been evaluated and the contributions of the new findings to current studies on radiation shielding have been emphasized.

Literature Review

Radiation is broadly divided into two categories: Non-ionizing and ionizing radiations. Since ultraviolet, infrared, microwaves, radio waves, and sunlight cannot capable to eject electrons from atoms, they are classified as non-ionizing radiation (Oto, Yıldız, Akdemir, & Kavaz, 2015). On the other hand, X-rays, gamma-rays, along with the α, β and neutron particles have enough energy to ionize an atom (Harish, Nagaiah, & Kumar, 2012). Due to this reason, they are characterized as ionizing radiation and they have extremely negative effects on human health such as cell mutation, organ damage, cancer, and even death (Elgazzar & Kazem, 2015).

As is known, ionizing radiation is mostly used both in nuclear facilities and in medicine for diagnosis and treatment purposes. The negative effects of ionizing radiation on human health vary depending on the exposure time, exposure amount, and the person is how close to the radiation source. To minimize the harmful effect of radiation, **A**s Low **A**s **R**easonably **A**chievable-ALARA principle -a guiding protocol for radiation safety- has been announced. According to ALARA protocol, to minimize the radiation risk on human health, the exposure time and the exposure amount should be reduced as much as possible and effective protectors should be used (Walker & Wellock, 2010).

Since it is not possible to avoid ionizing radiation such as X-rays and gamma-rays in the mentioned medical and nuclear facilities; the most effective protection from radiation is the use of armored clothing and equipment. From this point of view, materials with high atomic numbers (*Z*), their oxides, concrete, and steel have been used as shielding materials for radiation protection for many years. For example, while the ionizing radiation sources are kept in rooms with thick concrete walls reinforced with some special elements, aprons made of lead and its derivatives are designed for those working close to these devices. However, the significant disadvantages of these traditional protective materials have led researchers to seek new materials that can overcome these disadvantages. At the top of these disadvantages, the toxic effect of lead on the environment and human health can be expressed (AbuAlRoos, Amin, & Zainon, 2019). In addition, the heavyweight of the protective clothing made of materials such as lead and steel prevents them from being worn for long hours.

In the last decade, many materials and different material combinations known as composites are in a great race to replace lead and lead-derived materials in radiation shielding. In accordance with the usage area such as glasses, aprons, helmets, etc. many kinds of glass, glassy structure, and polymer composite are investigated for protection from radiation. Among these material families, polymers have unique properties such as low cost, low weight, stability over a wide temperature range, and high processability (Alavian & Tavakoli-Anbaran, 2019) that cannot be reached by lead. On the other hand, polymers alone cannot shield ionizing radiation such as X-rays or gamma rays, since most polymers are composed of the low atomic number elements carbon, oxygen, hydrogen, and nitrogen. To overcome this problem, polymers are usually reinforced with high *Z*-elements and their derivatives. Thus, a synergy is created by combining the lightness, flexibility, transparency, and environmental friendliness of the polymers with the high radiation shielding capabilities of the high-*Z* elements (bismuth, tungsten, barium, cerium, tin, etc.) (La et al. 2016; Szajerski, Zaborski, Bem, Baryn, & Kusiak, 2014) and the compounds containing those elements. Among these reinforcing materials, heavy metal oxides such as WO_3 and Bi_2O_3 are the most utilized particles to fabricate lead-free radiation shielding composites (Karabul & İçelli, 2021). Sometimes micro and sometimes nano-sized particles are preferred as reinforcing particles in the composite production stage. Although the increased surface-to-volume ratio for nanoparticle addition leads to more effective radiation performance (Kumar, Depan, Tomer, & Singh, 2009), the usage of nanoparticles increases the synthesis costs as well. In this regard, some remarkable results obtained from polymer-based micro and nanostructured composites that can act as radiation shields are as follows. In this short review, both experimental and theoretical studies performed by Monte Carlo N-Particle (MCNP) simulation and also WinXCom software have been given.

Very recently, Mahmoud et al. studied polyvinylchloride (PVC) composites in which chalconite and hematite minerals have been used as reinforcing particles. The WinXCom and MCNP5 based theoretical study results showed that the radiation attenuation ability of PVC can be increased by the addition of both minerals. However, the best improvement on the linear attenuation coefficient was recorded for the chalcotine doping between 0.015-15 MeV photon energies (Mahmoud, Lacomme, Sayyed, Özpolat, & Tashlykov, 2020). On the other hand, Avcıoğlu et al. investigated the radiation shielding properties of the low-density polyethylene (LDPE)/B_4C composites to get lightweight, eco-friendly, and flexible neutron shielding composites. Their results indicated that up to 39% B4C nano/sub-micro particle addition in weight increases the neutron shielding performance of LDPE efficiently (Avcıoğlu et al. 2020). Also, Nambiar et al. studied the X-ray attenuation performance of the polydimethylsiloxane/Bismuth oxide (PDMS/Bi_2O_3)

nanocomposites. They showed that the nanocomposite having 44.44% Bi_2O_3 nanoparticle has the best radiation shielding among other composites (Nambiar, Osei, & Yeow, 2013). Moreover, in another study isophthalic polyester (PES) has been reinforced with hematite by Belgin et al. to enhance the radiation shielding ability of PES. Their results indicate a promising alternative to lead since the lightweight PES/ 50 % Hematite composite has the mass attenuation coefficient of 0.098 cm^2g^{-1} against considerably high-energy gamma rays (662 keV) that is very close to the mass attenuation coefficient of lead for the same photon energy(Belgin et al. 2015). On the other hand, when the ionizing radiation shielding performances of some polymers are compared, Mann et al. reported that PVC exhibits the best gamma attenuation between 10 keV and 1400 keV. Moreover, Afhsar et al. tried to increase the radiation shielding ability of high-density polyethylene (HDPE) by doping it with different inorganic micro-sized particles such as W, B_4C, and MoS_2. Among these particles, W addition resulted in the highest attenuation against X-ray as well as desired flexibility (Afshar, Morshedian, & Ahmadi, 2019). Also, poly (methyl methacrylate)/bismuth oxide (PMMA/Bi_2O_3) composites have been tested for shielding of ionizing radiation by Cao et al. They showed that PMMA/44 %Bi_2O_3 composite has the highest mass attenuation coefficient for the low energy gamma rays (Cao, Yang, Bourham, & Moneghan, 2020). Furthermore, Korkut et al. tried to enhance the X-ray, gamma-ray, and neutron shielding ability of epoxy by doping it with FeCr slag. Their results were promising for 50% FeCr doping (Korkut, Gencel, Kam, & Brostow, 2013). Additionally, the WO_3 micro and nano-sized particles have been used to increase the radiation shielding ability of epoxy by Noor Azman et al. They indicated that the addition of nano-sized WO_3 particles leads to better radiation shielding against X-rays (Azman, Siddiqui, Hart, & Low, 2013). In addition to the studies mentioned so far, Mori et al. used a new ceramic raw material mostly composed of Fe_2O_3 for the protection from ionizing radiation. This new material had a mass attenuation coefficient which is approximately 40% of lead and twice that of concrete (Mori, Oku, Mannami, & Kunisada, 2020). These promising results indicate that such ceramic materials can be a good alternative for lead or concrete soon.

In addition to these studies, few studies have been carried out by using PP as the host matrix in producing radiation shielding composite materials. In 2019, Mirzaei et al. tried to produce lightweight, comfortable and low-cost textile-based protective clothes against X-ray. They fabricated garments woven with melt-spun polypropylene monofilaments. The results showed that the fabrics can be an appropriate barrier for X-rays utilized in medical applications (Mirzaei, Zarrebini, Shirani, Shanbeh, & Borhani, 2019). Furthermore, El-Sharkawy et al. studied the gamma-ray shielding properties of polypropylene ternary nanocomposites consisting of nano-bentonite and nanocadmium oxide particles. They concluded that the addition of 40% nanocadmium oxide particles increases the radiation shielding parameter of PP including mass attenuation coefficient, effective atomic number, etc. more than nano-bentonite (El-Sharkawy, Allam, El-Taher, Shaaban, & Mahmoud, 2021).

Based on the literature mentioned so far, this study has been focused on producing two kinds of PP-based composite materials. In the first composite series, Yahyalı stone (YS) powder mainly composed of Fe_2O_3 microparticles has been utilized as reinforcing particles to enhance the radiation shielding ability of the PP matrix. In the second PP-based composite series, Bi-based copper oxide layered ceramic material ($Bi_2Sr_2Ca_2Cu_3O_{10+x}$ shortly coded as Bi-2223) that includes high *Z*-elements have been used for the first time as the guest particles for the composites.

Theoretical Background

The materials to be utilized in radiation protection should have homogeneous density and content as well as sufficient thickness to reduce the radiation dose to a safety level. Some internationally accepted parameters show the performance of a material produced with such properties to shield the incoming radiation. These parameters are linear attenuation coefficient (μ), mass attenuation coefficient (μ/ρ), mean free path (MFP), half-value thickness (HVL), tenth-value layer (TVL), effective atomic number (Z_{eff}), and effective electron density (N_{eff}).

As is known, when radiation passes through a material environment, it reduces its intensity by making possible material-photon interactions including pair production, Compton scattering, and photoelectric effect, etc. Based on experimental observations, it is known that the decrease in incoming radiation intensity is exponential due to material-photon interactions. The decrease in intensity of an incoming beam obeys Beer-Lambert law:

$$I(x) = I_0 \exp(-\mu x) \tag{1}$$

where I_o and I are the intensity of the incoming and the transmitted ray, respectively. According to Eq. (1), the decrease in the intensity of the radiation is directly proportional to the thickness of the shielding material, x. The linear attenuation coefficient used as the radiation interaction parameter doe not depend on the phase of the shielding material. Depending on the mass density (ρ) of the radiation-absorbing medium, the mass attenuation coefficient is defined. μ/ρ also represents the total probability of interaction of a shielding material with a monoenergetic photon per unit volume and has the unit of cm^2/g.

HVL and TVL are the material thicknesses required to reduce the incoming radiation intensity to half and one-tenth, respectively. In this context, these critical lengths are described by using Eq. (1)

$$HVL = \frac{\ln 2}{\mu} \tag{2}$$

$$TVL = \frac{\ln 10}{\mu} \tag{3}$$

Another critical length is MFP that describes the average distance between the two consecutive interactions between the radiation and matter through the matter:

$$MFP = \frac{1}{\mu} \tag{4}$$

When a radiation absorbing material is composed of at least two elements, the Z number of the material is defined by using a single number known as Z_{eff} that is also another important interaction parameter between photon and matter. (Hine, 1952; Petrucci, 2017). Z_{eff} is determined by the interpolation method and given in Eq. (5):

$$Z_{eff} = \frac{Z_1(log\sigma_2 - log\sigma_{sample}) + Z_2(log\sigma_{sample} - log\sigma_1)}{log\sigma_2 - log\sigma_1} \tag{5}$$

where σ_{sample} is the total atomic cross-section for each sample (Çağlar et al. 2019). and given in Eq. (6):

$$\sigma_{sample} = \frac{(\mu/\rho)_{sample}}{N_A \sum_i f_i / A_i} \tag{6}$$

In Eq. (6), A_i and f_i are the atomic mass and molar fraction of the ith element, respectively and N_A is the Avogadro constant.

The σ of each element present in the specimen is calculated for obtaining the σ_1 and σ_2. When σ_{sample} is compared with the σ of each element present in the absorbing material, σ_1 is equal to the σ of the lowest value near σ_{sample} whereas σ_2 corresponds to the highest value near σ_{sample}. Consequently, Z_1 and Z_2 in Eq. (5) also corresponds to the Z values of the elements with the cross-sections of σ_1 and σ_2, respectively.

As the last important radiation shielding parameter, effective electron density (N_{eff}) describes the photon-matter interaction. N_{eff} is defined by using Eq. (7).

$$N_{eff} = N_A \frac{Z_{eff}}{<A>} \tag{7}$$

where $< A >$ is the average atomic mass defined by

$$< A >= \sum_i \frac{n_i A_i}{n} \quad (8)$$

According to Eq. (7), N_{eff}, which is the number of electrons per unit mass of the radiation shielding material, is directly proportional to the Z_{eff} (Manohara, Hanagodimath, Thind, & Gerward, 2008).

MATERIALS, CHARACTERIZATION AND METHODS

Materials

In this study, two different composite families based on PP have been prepared by using different raw materials. The PP granules with a technical grade have been supplied from Petkim Industry (Turkey). The molecular weight of the PP is 42.08 g/mol, its melting temperature is 165°C, and the mass density of PP is 0.905 g/cm^3.

While YS which is a natural stone collected from Kayseri, Tukey has been used for the PP/YS composites, Bi-2223 ceramic fine powder has been produced for the PP/Bi-2223 composites. For the synthesis of Bi-2223 ceramic, Bi_2O_3, SrO, CaO, and CuO fine powders with a purity greater than 99.99 % have been purchased from Sigma Aldrich. The code numbers of the fine powders are 202827 for Bi_2O_3, 415138 for SrO, 451711 for CaO, and 203130 for CuO.

The Reinforcing Ceramic Powder: Preparation and Chemical Composition

Bi-2223 ceramic fine powder has been produced by the solid-state reaction method. The high purity oxide powders including Bi_2O_3, SrO, CaO, and CuO have been weighted appropriate amount of mass to prepare $Bi_2Sr_2Ca_2Cu_3O_{10+x}$ ceramic and the powders have been mixed in an agate mortar for 2 hours. The mixture has been pressed under approximately 10^9 Pa pressure to obtain a pellet form. Then the pellet has been put into an aluminum oxide boat and heated at a rate of 1°C/min from room temperature to 900°C and then kept in the oven at 900°C for 24 h in normal atmospheric pressure and environment. The pellet has been kept in the oven for one day to cool and then it has been ground in an agate mortar again. After grinding, the powder has become a pellet, and then a sintering process has been performed at 950°C for 24 h. After the sintering process, the resulting pellet has been ground finally to obtain a fine powder consisting of ceramic micro-particles. The chemical content of the ceramic powder has also been determined by X-Ray Fluorescence (XRF) spectrometer with the Axios-Advanced (PANalytical B.V.) and shown in Figure 1.

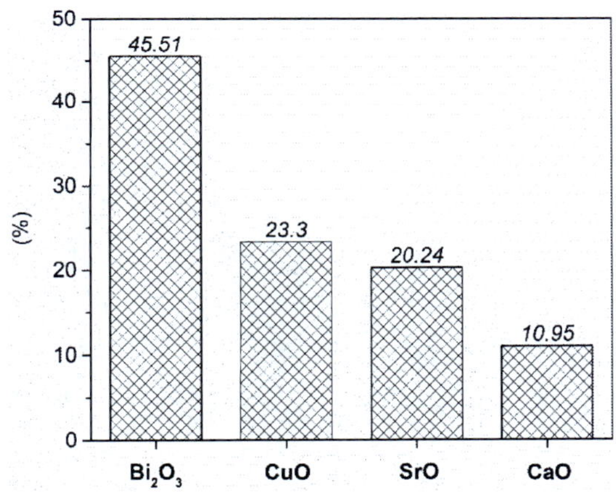

Figure 1. Chemical composition of ceramic powder.

The Reinforcing Natural YS: Chemical Composition

The chemical content of the YS powder has also been given in Figure 2. According to Figure 2, YS powder appears to contain a large amount of Fe_2O_3 at a high rate of 75.27%. The other most abundant compound in YS is SiO_2 (17.21%), which is also in the content of most natural stones. SiO_2 is also followed by Al_2O_3 with 4.24%.

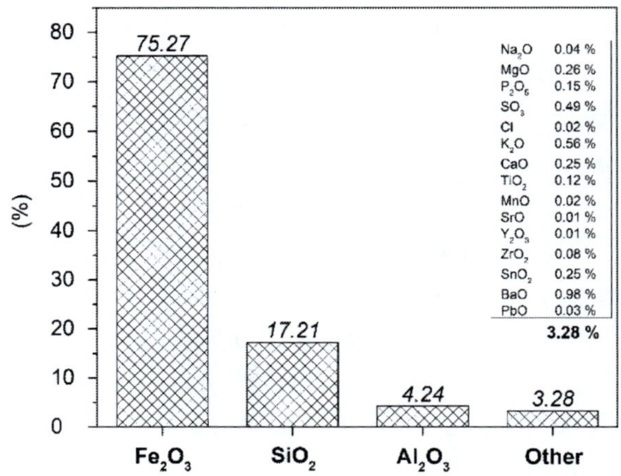

Figure 2. Chemical composition of YS.

Preparation of the Composites

The two composite families have been prepared by using a single screw extruder. First of all, PP in the form of granules has been pulverized with a grinder. To prepare PP-based composites, reinforcing powder either YS or Bi-2223 has been added to PP powder in different weight percentages (5%, 10%, 20%, and 30%) Then the mixtures have been processed in the single screw extruder twice to obtain homogenous mixtures. Afterward, the samples have been put into a circular Teflon mold and a hot press at 165-175 °C has been applied. The mold has a diameter of 20 mm and the thickness of the samples were varying between 8 mm and 12 mm. In addition to these composites, as a reference, a pristine PP sample has been prepared by using the same procedure mentioned above. Additionally, the mass density values of the samples have been calculated by using the Archimedes principle and are listed in Table 1.

As is seen from Table 1, the mass density of the samples increases with the raising additive weight percentage. Moreover, the highest mass density value has been observed for the PP/ 30% YS composite.

Table 1. The experimental mass density values of the samples

Materials	Density (g/cm^{-3})	Materials	Density (g/cm^{-3})
PP	0.92		
PP/ 5% Bi-2223	0.98	PP/ 5% YS	0.95
PP/ 10% Bi-2223	1.02	PP/ 10% YS	1.05
PP/ 20% Bi-2223	1.07	PP/ 20% YS	1.06
PP/ 30% Bi-2223	1.16	PP/ 30% YS	1.24

Methods

The gamma-ray shielding performances of the samples have been determined experimentally and theoretically. The details of the experimental setup and the WinXCom software which has been used for theoretical calculations have been introduced as follows.

Gamma-ray spectroscopy has been used in the experimental radiation shielding parameters. As is known, gamma spectroscopy is a method generally used to determine the radioactivity of material and the density of radioisotopes. The spectroscopy system consists of a detector, electronic systems connected to it (such as photomultiplier tube, amplifiers analog-to-digital signal converter, and multi-channel analyzer), and a computer with the software installed. The main goal of spectroscopy is to analyze the peaks in the spectrum for signals corresponding to energy absorption, which are generated by processing the signals from the detector. The system is surrounded by various protective collimators and shields to prevent unwanted noises and counts that may occur while measuring. While a schematic representation of gamma-ray spectroscopy has been given in Figure 3(a), Figure 3(b) shows the geometry and other details of the experimental setup.

The working principle of the NaI(Tl) detector used in the experiments is the same as in a typical scintillator detector. The thallium-activated sodium iodide detector reacts to gamma-rays by creating a light flash or a scintillation. When the excited electrons return to their ground state, a scintillation occurs. The detector crystal is mounted in a photomultiplier tube (PMT) that is responsible for converting the scintillation mentioned into an electrical pulse. The initial signal of the photocathode is very small, and the signal is amplified by a series of dynodes to obtain a large signal. The NaI(TI) scintillation detector, ORTEC brand 905-4 model, is 3×3 inches in size. As shown in Figure 3(b), there is an ORTEC 276 model preamplifier connected to the back of the relevant detector. The signal coming out of the preamplifier comes to the ORTEC 672 model main amplifier, where the signal is amplified and the signal shaping is done. The signals coming from the amplifier come to the interface card consisting of 2041 channels in the computer via an analog-digital signal converter (ADC) and then observed as a spectrum on the ORTEC brand 927 ASPEC MCA model multichannel analyzer screen. The analyzer is controlled by a computer software program called MAESTRO. The PMT, which is connected to the detector system, is digiBASE (Ortec) and has a diameter of 6.3 cm and a length of 8.0 cm. The PMT is separated from the *N*aI(Tl) crystal by a 5 mm thick glass window.

As radioactive point sources, Ba-133, Cs-137, and Co-60 have been chosen to determine the gamma-ray shielding capabilities of the composites produced. The gamma photons most likely to be emitted from the Ba-133 source have energies of 81 keV, 303 keV, 356 keV, and 384 keV. The experimental study has been carried out for photons with 81 keV and 356 keV energies for Ba-133. On the other hand, experiments have also been performed for the photon energies of 662 keV emitted from the Cs-137 source and the 1173 keV and 1332 keV, which are the most likely to be emitted from the Co-60 source.

On the other hand, the theoretical calculations for the radiation shielding parameters have been performed by using WinXCom software that serves as a very practical and reliable option to obtain the linear attenuation coefficient as well as cross-sections for compounds and mixtures against ionizing radiation. The software has been developed by Gerward et al. at the beginning of the 21st century (Gerward, Guilbert, Jensen, & Levring, 2001).

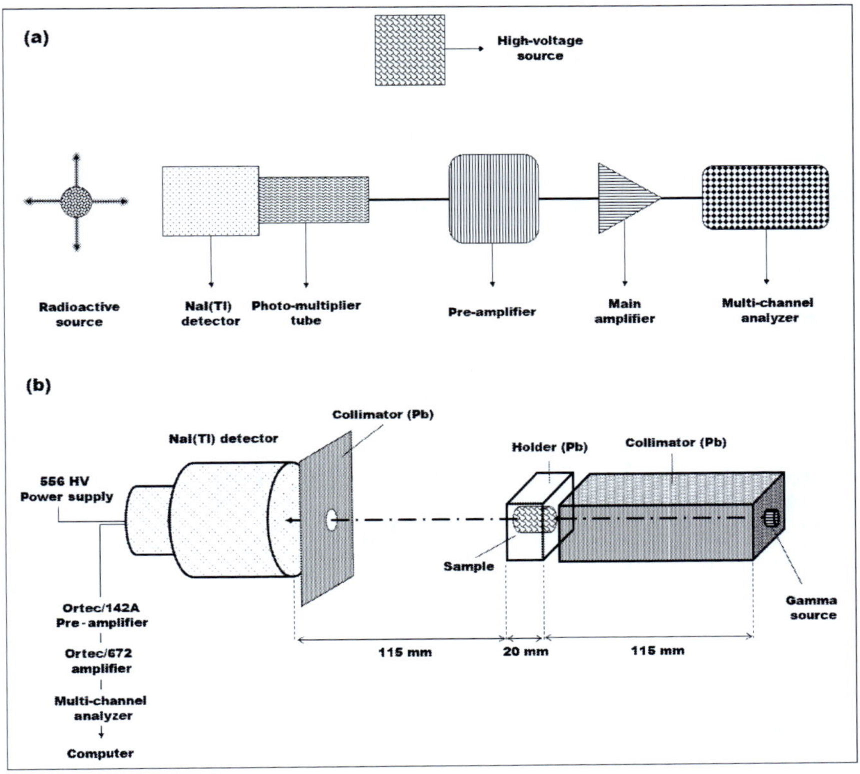

Figure 3. (a) The schematic representation of gamma-spectroscopy and (b) Schematic representation of the shielding measurement geometry (Şahin, Bozkurt, Karabul, Kılıç, & Özdemir, 2021).

Results and Discussion

This section has been devoted to investigate and interpret the role of ceramic and natural stone addition on the radiation shielding ability of PP in the context of linear attenuation coefficient and the critical lengths of HVL and MFP. Along with this study, matter-photon interactions, which are effective at different photon energies, have also been mentioned. On the other hand, Z_{eff} and N_{eff} have been interpreted with the help of both WinXCom and experimental data in a wide photon energy range. Moreover, the radiation shielding potential of the samples prepared in this study has been tried to reveal by comparing the results with each other and with the current studies from the scientific literature. In this context, the results and discussion section has been divided into two subsections as follows:

Linear Attenuation Coefficient and Critical Lengths of the Samples

This section has discussed how well all samples, including pristine PP, can shield the radiation corresponding to photons of different energies emitted from 3 point radioactive sources. In this context, Table 2 shows the experimental and WinXCom software-derived μ values of both composite families for the photons having 81 keV and 356 keV energies.

First of all, when the μ values of the samples against photons with 81 keV energy are evaluated, it can be stated that both additives (Bi-2223 and YS) improve the radiation shielding of the pristine PP by increasing the μ value. Additionally, both Bi-2223 and YS increase the mass attenuation coefficient hierarchically as their contribution percentage raises in the composite. In the context of experimental data, when the additive particle weight percentage reaches its maximum value ie. 30%, the increment of the linear attenuation coefficient has been determined as approximately 4.48 and 1.87 times for the Bi-2223 and YS, respectively. On the other hand, when the same Bi-2223 and YS additive percentages are compared, the reason why higher μ values are observed in PP-based composites with Bi-2223 additive is due to the fact that the Bi-2223 ceramic additive contains elements with higher atomic numbers than the YS additive content. In addition, observation of the lowest μ value for the pristine PP among all samples can be explained by the fact that its content consists of elements with low atomic numbers such as carbon and hydrogen.

When the significant increase in the linear attenuation coefficient observed with increasing additive percentage against 81 keV photons for both composite families is considered in the context of fundamental photon-matter interactions, it is seen that the photoelectric effect is the dominant interaction mechanism at photon energies as low as 81 keV. Because, as can be seen from Eq. (9), the cross-section of photoelectric interaction ($\sigma_{photoelectric}$) is highest for the low energetic photons.

$$\sigma_{photoelectric} \propto \frac{1}{E^{3.5}} \tag{9}$$

where E is the photon energy. Moreover, as was reported by Berger et al. attenuation of radiation by photoelectric interaction constitutes the largest contribution to the total cross-section at low photon energies for the absorbing materials with high Z elements (M. Berger, 2010). In this respect, the fact that the highest μ value is observed in the contribution of 30% Bi-2223 instead of 30% YS is due to the higher number of photoelectric interactions occurring in the contribution of Bi-2223 with higher Z elements content than YS.

Table 2. The experimental and WinXCom derived linear attenuation coefficients (cm^{-1}) of the PP-based Bi-2223 and YS enriched samples against the photons emitted from Ba-133 point radioactive source

	81 keV		**356 keV**	
	Experimental	WinXCom	Experimental	WinXCom
PP	0.171	0.167	0.105	0.105
PP/ 5% Bi-2223	0.260	0.240	0.118	0.115
PP/ 10% Bi-2223	0.363	0.311	0.137	0.122
PP/ 20% Bi-2223	0.574	0.468	0.155	0.132
PP/ 30% Bi-2223	0.766	0.658	0.175	0.158
PP/ 5% YS	0.209	0.196	0.115	0.114
PP/ 10% YS	0.233	0.218	0.123	0.119
PP/ 20% YS	0.286	0.259	0.132	0.124
PP/ 30% YS	0.317	0.320	0.135	0.136

The same trend has also been observed for the linear attenuation coefficient of the samples against the photons having 356 keV energy (See Table 2). However, since the photon energy is increased to 356 keV, the μ values were even lower than the values observed for the same additive contribution percentage at 81 keV. This response of the samples indicates a decrease in the ability of the samples to attenuate the radiation with increasing photon energy. The same response can also be clarified by the variation of the predominant interaction mechanism between the matter and photon from the photoelectric effect to Compton scattering. Since Compton scattering cross-section is proportional to incident photon energy as follows (El-Agawany, Mahmoud, Kavaz, El-Mallawany, & Rammah, 2020),

$$\sigma_{Compton} \propto \frac{1}{E} \tag{10}$$

The possibility of observing photoelectric interaction decreases as the photon energy increases and the probability of occurrence of Compton scattering increases simultaneously. Furthermore, Bichsel and Schindler emphasized that the possibility of observing Compton scattering is high for the mid-photon energies and the target materials having elements with atomic numbers between $Z = 1$ and $Z = 60$ (Bichsel & Schindler, 2020).

When all the data in Table 2 are considered together, it has been determined that especially PP/30%Bi-2223 composite shields the ionizing radiation better than all composites against photons with 81 keV and 356 keV energies. Besides, Table 2 can be compared in terms of experimental and WinXCom data. As a result of this comparison, it is seen that there is an agreement between experimental and WinXCom findings in terms of increasing μ values with increasing additive percentage and in terms of the values of μ. Following these findings, mass attenuation coefficients of the samples, calculated from the μ values in Table 2 by using the mass densities given in Table 1, can be compared with some radiation shielding materials: Aluminum and Copper. As is known, aluminum and copper have mass attenuation coefficients of

0.2215 cm^2g^{-1} and 0.7946 cm^2g^{-1} for 81 keV photons, respectively (Gjorgieva & Barandovski, 2016). On the other hand, 20% Bi-2223 and 30% Bi-2223 doped PP composites have the experimental mass attenuation coefficients of 0.530 cm^2g^{-1} and 0.658 cm^2g^{-1} at 81 keV, which are significantly greater than that of aluminum and comparable with the copper for the same photon energy. Moreover, the PP/30% Bi-2223 composites have a higher mass attenuation coefficient than other polymer-based microstructured composites. Isophthalate Polyester/WO_3 composite having the mass density of 1.25 gcm^{-3} exhibited radiation attenuation against 81 keV photons with the mass attenuation coefficient that is lower than 0.500 cm^2g^{-1} (Ambika, Nagaiah, & Suman, 2017). Also the PP/30% Bi-2223 composite has the mass attenuation coefficient of 0.148 cm^2g^{-1} against 356 keV photons. If the mass attenuation coefficients of the PP/30% Bi-2223 composite for 81 keV and 356 keV are compared with the stainless steel SS304L (0.587 cm^2g^{-1} at 81 keV and 0.099 cm^2g^{-1} at 356 keV) (M. J. Berger & Hubbell, 1987), it is seen that the composite appears to have a radiation shielding advantage over stainless steel at the respective energies. Moreover, the PP/30% Bi-2223 composite has a better radiation shielding performance against 81 keV and 356 keV photons than a composite composed of continuous basalt fiber, epoxy, and erbium oxide having the mass attenuation coefficients of 0.407 cm^2g^{-1} at 81 keV and 0.104 cm^2g^{-1} at 356 keV (Li, Gu, Zhang, et al. 2017). Furthermore, the PP-based composite consisting of the highest Bi-2223 ceramic content has a competitive performance against the photons having 356 keV energy with the 20.1 wt% nano-Gd_2O_3 enriched epoxy polymer composite since its mass attenuation coefficient was reported as 0.103 cm^2g^{-1} by Li et al. (Li, Gu, Wang, et al. 2017).

To point out the radiation shielding capabilities of the samples prepared, the results of the experimental and theoretical studies for the critical lengths of HVL and MFP are given in Figure 4.

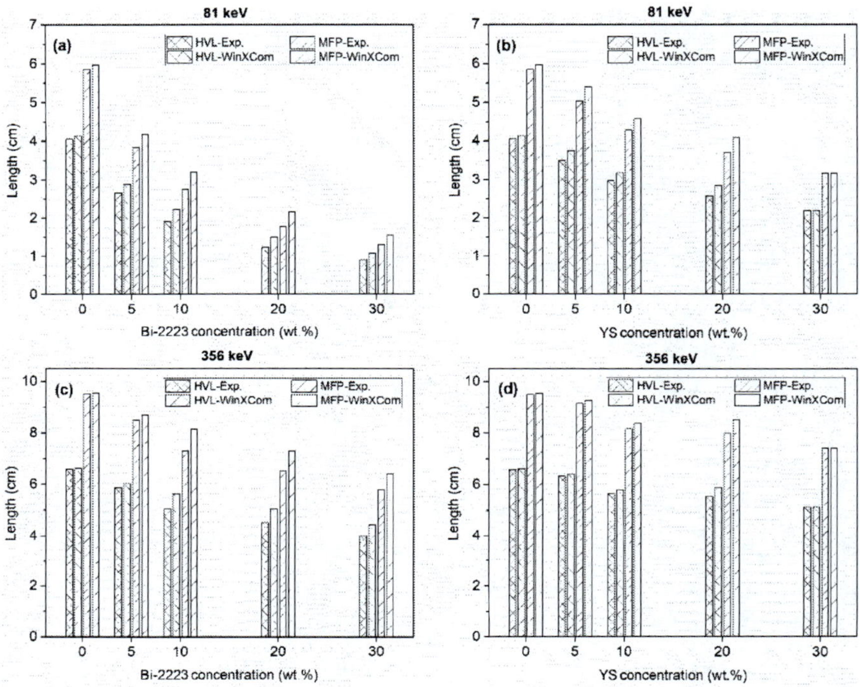

Figure 3. Variation of the experimental and WinXCom derived HVL and MFP values of the PP with increasing additive percentages of Bi-2223 at (a) 81 keV and (c) 356 keV and for YS at (b) 81 keV and (d) 356 keV.

In accordance with the determinations made for linear attenuation coefficient values, decreasing critical length values with the increasing percentage of additive particles have been determined in the HVL and MFP values, whose changes are given in cm (See Figure 4). This result is valid for both photon energies (81 keV and 356 keV) and both contributions (Bi-2223 and YS). However, it has been observed that the most effective ionizing radiation shielding, namely the shortest HVL and MFP values, is provided for the PP-based composite containing the highest concentration of ceramic additives against photons with 81 keV energy.

If the average values of the experimental and theoretical HVL values for 81 keV in Figure 4 are calculated for the pristine PP and the composites containing the highest concentration of reinforcing particles, while the average value of HVL of the pristine PP was 4.099 cm; $HVL_{average}$ values for the PP/30% Bi-2223 and PP/30%YS composites were 0.985 cm and 2.189 cm, respectively. In this case, the HVL value of the pristine PP decreases by 4.16 and 1.87 times for the 30% Bi-2223 and 30% YS contributions, respectively. If the analysis is extended to 356 keV, as expected, the shortening in the average HVL values with the additive slowed down, and the shortening in the average HVL was approximately 1.56 and 1.28 times for the maximum Bi-2223 and YS additives, respectively. Continuing with the comparison of some HVL data in Figure 4 with some studies in the scientific literature, for example, the experimentally determined HVL value for the 30% Bi-2223 doped PP composite at 81 keV (0.904 cm) is lower than that of the Isophthalic polyester/Hematite composite (HVL = 0.936 cm at 88 keV) prepared by Belgin et al. (Belgin et al. 2015). Furthermore, the HVL values of 4.02 cm and 5.15 cm for the PP/30% Bi-2223 and PP/30% YS composites are both less than that of High-Density polyethylene /10 wt% nano CdO composite (HVL = 5.296 cm) against 356 keV photons (El-Khatib et al. 2019). On the other hand, if Figure 4 is evaluated in terms of MFP, the shortened MFP values can be attributed to the increased interaction between matter and photons.

In Table 3, the linear attenuation coefficients of the two composite families have been shown for the photons having 662 keV energy emitted from the Cs-137 point radioactive source. Similar to the trend observed for 81 keV and 356 keV energy photons, the linear attenuation coefficient increases with increasing Bi-2223 ad YS additives but the speed of increase is low relative to the low energy photon case. When the μ values are compared with each other, it has been confirmed from both experimental and WinXCom data that the Bi-2223 additive is more successful than the YS additive in shielding the photons with 662 keV energy. The increments in μ value have been determined as 1.40 times and 1.27 times relative to pristine PP for Bi-2223 and YS additives, respectively. This behavior can be explained by the dominant Compton scattering mechanism for the attenuation of radiation. Therefore, the linear attenuation coefficient for 662 keV cannot reach the high values observed for 81 keV or 356 keV.

Table 3. The experimental and WinXCom derived linear attenuation coefficients (cm^{-1}) of the PP-based Bi-2223 and YS enriched samples against the photons emitted from Cs-137 point radioactive source

	662 keV	
	Experimental	WinXCom
PP	0.079	0.081
PP/ 5% Bi-2223	0.088	0.087
PP/ 10% Bi-2223	0.100	0.090
PP/ 20% Bi-2223	0.107	0.094
PP/ 30% Bi-2223	0.111	0.103
PP/ 5% YS	0.087	0.083
PP/ 10% YS	0.093	0.092
PP/ 20% YS	0.095	0.090
PP/ 30% YS	0.100	0.104

According to Table 3, the PP/30% Bi-2223 and PP/30% YS composites have better radiation shielding ability against 662 keV than the Epoxy/15% Fe_3O_4 nanocomposite, since the nano iron oxide particle doped epoxy had the linear attenuation coefficient of 0.0971 cm^{-1} (Kathem, Resen, & Jubier, 2018). Additionally, the PP/30% Bi-2223 and PP/30% YS composites have both greater linear attenuation coefficients at 662 keV than another polymer-based composite consisting of poly(hydroxyethyl methacrylate) (PHEMA)/10% tungsten(VI) oxide (μ = 0.098 cm^{-1}) (Körpınar, Öztürk, Çam, & Akat, 2020). If the comparison is continued with the mass attenuation coefficient, the μ/ρ values of especially PP/30% Bi-2223 (μ/ρ = 0.095 cm^2g^{-1}) and PP/20% YS (μ/ρ = 0.090 cm^2g^{-1}) composites can be compared with other polymer-based composites. For example, both composites prepared in the present study have greater μ/ρ values than Epoxy/6% Fe_3O_4 (μ/ρ = 0.0827 cm^2g^{-1}) and Epoxy/15% Al_2O_3 (μ/ρ = 0.0827 cm^2g^{-1}) (Aldhuhaibat, Amana, Jubier, & Salim, 2021) composites, making them better radiation shielders for 662 keV photons. Additionally, the PP/30% Bi-2223 (μ/ρ = 0.095 cm^{2}g^{-1}) and PP/20% YS (μ/ρ = 0.090 cm^{2}g^{-1}) composites have better radiation shielding ability than 50% PVA/50% WO_3 composite for 662 keV since the 50% PVA/50% WO_3 composite has the μ/ρ = 0.087 cm^{2}g^{-1} (Muthamma, Gudennavar, & Gudennavar, 2020).

Considering the analysis in the context of critical lengths (See Figure 5), the decreasing HVL and MFP values are observed with increasing reinforcing particle percentage for both composite families. However, the decrease in critical lengths here is not as sharp as observed at 81 keV. It is an indication that these samples cannot shield ionizing radiation at 662 keV very successfully. On the other hand, the PP/30% Bi-2223 and PP/30% YS composites having the HVL values of 6.26 cm and 6.90 cm display more effective radiation shielding for 662 keV photons than 5% wt. and 10% wt. lead oxide enriched isophthalic resin composites. The HVL values of these lead oxide-filled composites were reported as greater than 8 cm by Harish et al. (Harish et al. 2012). Moreover, the PP/30% Bi-2223 composite is competitive with the polyester composite filled with CdTe up to 20% wt at 662 keV since the polyester-based composites have HVL values around 6 cm for 662 keV (Akman et al. 2021).

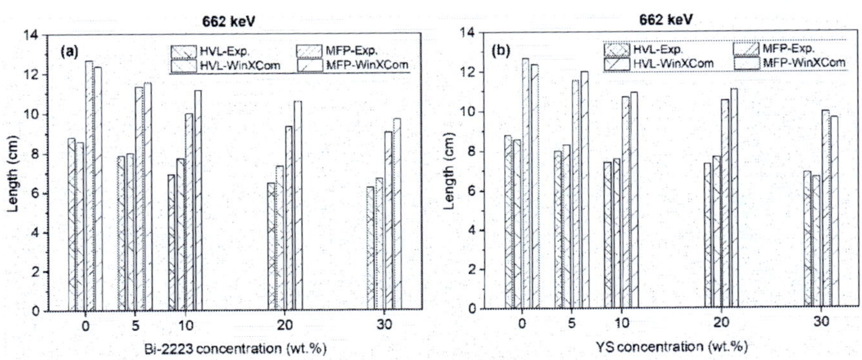

Figure 5. Variation of the experimental and WinXCom derived HVL and MFP values of the PP with increasing additive percentages of (a) Bi-2223 and (b) YS at 662 keV.

Finally, the radiation shielding performances of the two PP-based composite families have been determined experimentally and theoretically for the photons against the photons with 1173 keV and 1332 keV energies. Their radiation shielding capabilities have been evaluated by considering the variation of the linear attenuation coefficient and the critical lengths with the gradual increase of the reinforcing particle percentage shown in Table 4 and Figure 6.

Table 4. The experimental and WinXCom derived linear attenuation coefficients (cm^{-1}) of the PP-based Bi-2223 and YS enriched samples against the photons emitted from Co-60 point radioactive source

	1173 keV		**1332 keV**	
	Experimental	WinXCom	Experimental	WinXCom
PP	0.059	0.062	0.060	0.058
PP/ 5% Bi-2223	0.064	0.066	0.065	0.061
PP/ 10% Bi-2223	0.073	0.067	0.067	0.063
PP/ 20% Bi-2223	0.077	0.070	0.069	0.065
PP/ 30% Bi-2223	0.080	0.075	0.071	0.070
PP/ 5% YS	0.063	0.064	0.059	0.060
PP/ 10% YS	0.068	0.070	0.066	0.065
PP/ 20% YS	0.070	0.070	0.065	0.064
PP/ 30% YS	0.083	0.079	0.073	0.074

Considering the change of the radiation shielding parameters of the pristine PP with increasing additive percentages in Table 4 and Figure 6 in the context of both photon energies, even if the percentage of dopant particle was increased to the highest value, neither a significant increase in the linear attenuation coefficient nor a significant decrease in critical lengths has been obtained. This case can be interpreted that to attenuate the high energy ionizing radiation Bi-2223 and YS additives are not effective for the PP. On the other hand, it has been observed that at such high photon energies, the YS additive results in a slightly more successful radiation shielding than Bi-2223 for PP. Furthermore, although the samples do not exhibit very effective radiation shielding, they have competitive radiation shielding parameters with similar polymers and polymer-based composites. For instance, the PP-based composites having the highest percentages of Bi-2223 and YS additives exhibit better radiation shielding performance than rubber and jute fiber reinforced rubber composites since the linear attenuation coefficient of these materials were reported as less than 0.07 cm^{-1} for 1173 keV and 1332 keV photons (Parlar, Abdlhamed, & Akkurt, 2019). Similarly, the PP/30% YS composite has HVL parameters almost equal to the HVL values of PVA-based $BaTiO_3$ composites for 1173 keV and 1332 keV (Issa et al. 2021).

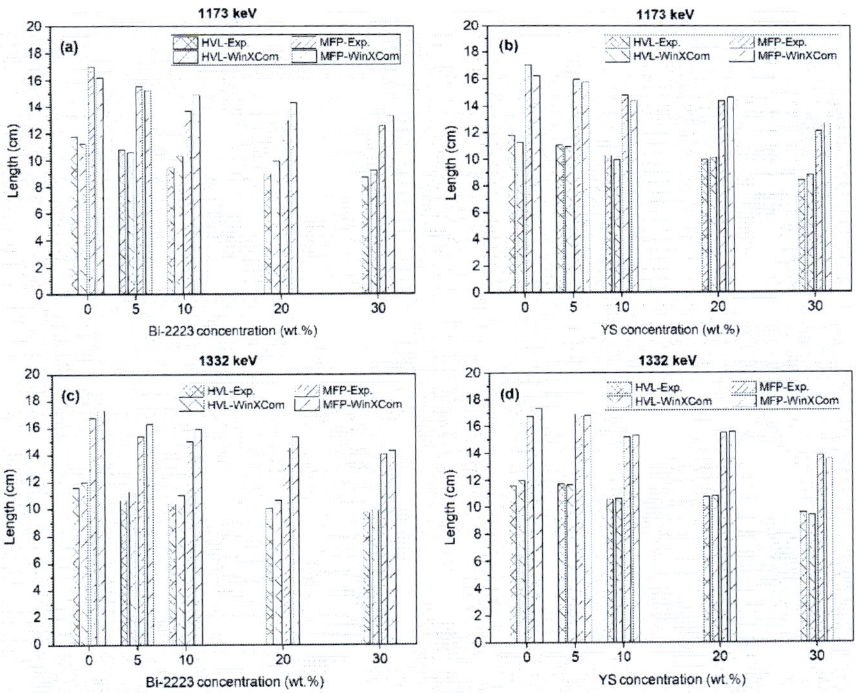

Figure 6. Variation of the experimental and WinXCom derived HVL and MFP values of the PP with increasing additive percentages of Bi-2223 at (a) 1173 keV and (c) 1332 keV and for YS at (b) 1173 keV and (d) 1332 keV.

Effective Atomic Number and Effective Electron Density of the Samples

Since the effective atomic number and effective electron density are useful parameters to determine the radiation response of composites ie. a multi-element material for both the industrial and technical implements, the second part of the results and discussion is devoted to analyze the variation of Z_{eff} and N_{eff} with respect to both incident photon energy and additive weight percentage. The analysis has been carried out both experimentally and theoretically by WinXCom software. From this point of view, while Z_{eff} and N_{eff} parameters have been determined experimentally for the photon energies of 81 keV, 356 keV, 662 keV, 1173 keV, and 1332 keV; the same parameters have been calculated by WinXCom software for 87 different photon energies varied from 1×10^{-3} MeV to 1×10^{5} MeV.

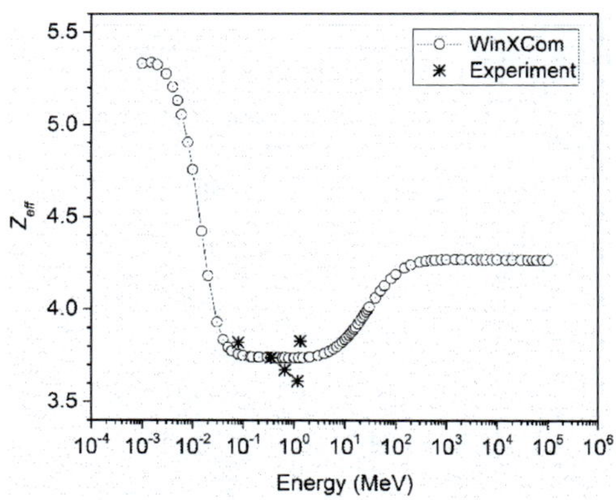

Figure 7. Variation of the experimental and WinXCom derived Z_{eff} values of the pristine PP with increasing photon energy.

The variations of the effective atomic number of the pristine PP, PP/Bi-2223, and PP/YS composites with increasing incoming photon energy have been shown in Figures 8, 9, and 10, respectively. As shown in the related figures, there is a good agreement between the experimental and WinXCom findings for the composites. On the other hand, some discrepancies have been observed between the Z_{eff} values determined experimentally and theoretically by WinXCom for the pristine PP, since pure PP does not contain any high atomic number element.

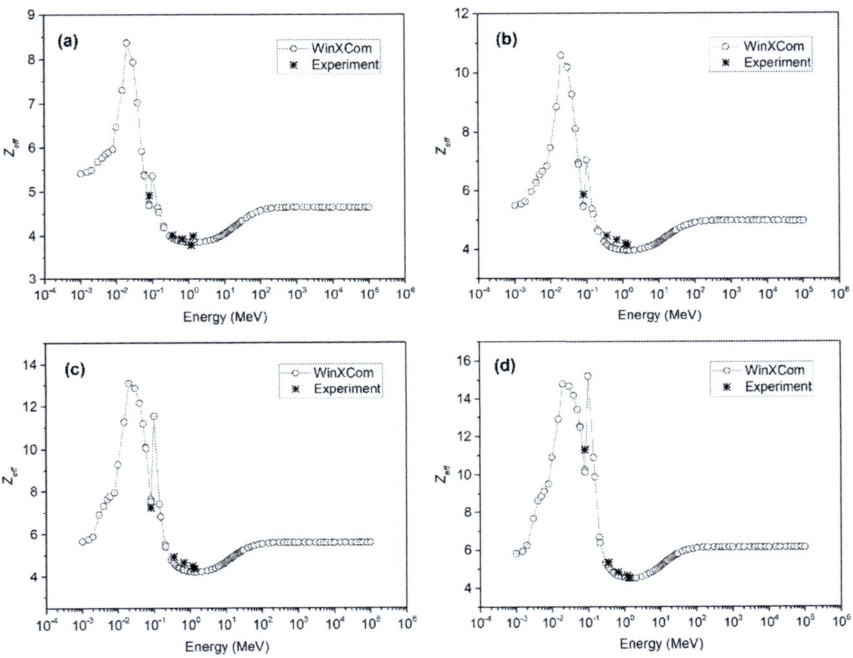

Figure 8. Variation of the experimental and WinXCom derived Z_{eff} values of the (a) PP/5% Bi-2223, (b) PP/10% Bi-2223, (c) PP/20% Bi-2223, and (d) PP/30% Bi-2223 composites with increasing photon energy.

The analysis can be started by interpreting which material-photon interaction mechanisms are dominant in which photon energy regions with the help of Z_{eff} versus photon energy graphs of all samples. It has been noted that all Z_{eff} versus photon energy curves have the same form. As shown in Figures 7, 8, and 9, up to a critical value of Z_{eff} depending on the absorbing material content, Z_{eff} exhibits an increasing trend with raising incoming photon energy and then decreases sharply until it reaches its lowest value. For the Bi-2223 doped PP-composites, during the decreasing of Z_{eff}, there is a sharp increase for one specific photon energy (0.1 MeV). The reason for this behavior can be explained by the K-edge discontinuity of bismuth observed at 0.0905 MeV (Ma, Holmes,

& Pourmorteza, 2019). After passing through its lowest value, Z_{eff} starts to increase slowly and display a saturation plateau for a very high photon energy range. These different behaviors of Z_{eff} for different photon energy ranges are of course due to the existence of different dominant photon-matter interaction mechanisms that are valid in different photon energy intervals. From this point of view, the Z_{eff} versus photon energy graph can be examined theoretically by dividing it into four regions as illustrated in Figure 10. In Figure 10, the numerical values of the Z_{eff} have been removed from the vertical axis to extend this analysis from the pristine PP to the PP-based composites prepared in the present study.

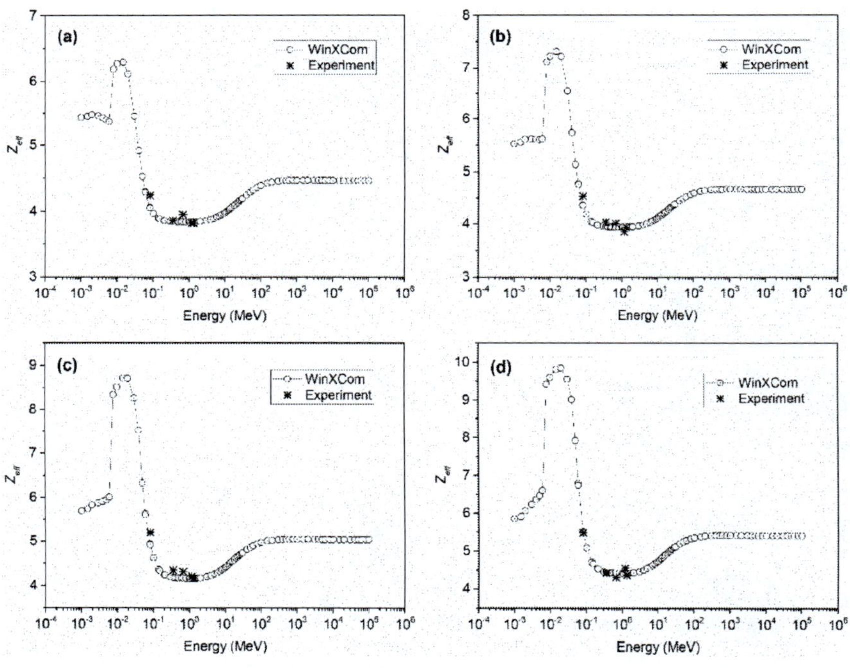

Figure 9. Variation of the experimental and WinXCom derived Z_{eff} values of the (a) PP/5% YS, (b) PP/10% YS, (c) PP/20% YS, and (d) PP/30% YS composites with increasing photon energy.

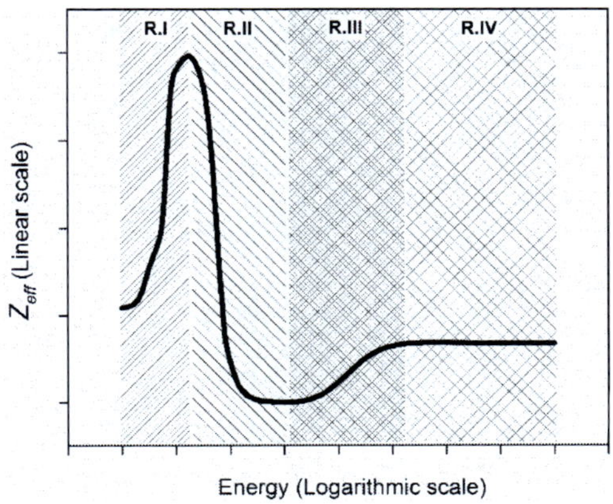

Figure 10. The general behavior of Z_{eff} with increasing photon energy (Şahin et al. 2021).

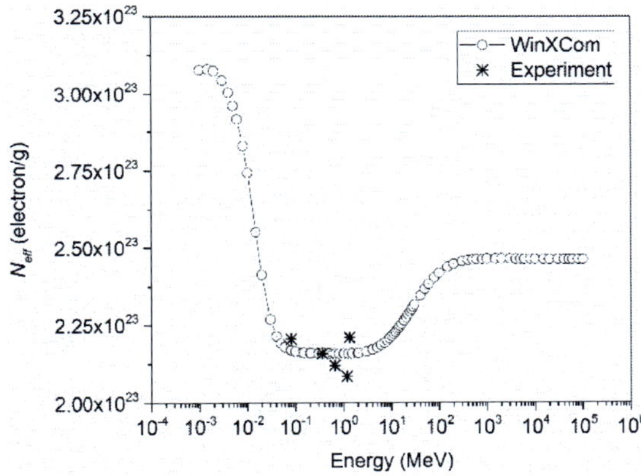

Figure 11. Variation of the experimental and WinXCom derived N_{eff} values of the pristine PP with increasing photon energy.

When Figure 10 is considered according to the dominant material-photon interaction mechanisms, while photoelectric absorption becomes the dominant mechanism, which manifests itself as an increase in Z_{eff} in the low photon energy region (Region I); the decrease observed in Z_{eff} in the middle photon energy region (Region II) mostly ascribes the dominant Compton scattering. The second increase in the Z_{eff} for the high photon energy interval (Region III) is attributed to the increased possibility of pair production. Finally, the dominant interaction mechanism is pair production for very high photon energies described in Region IV.

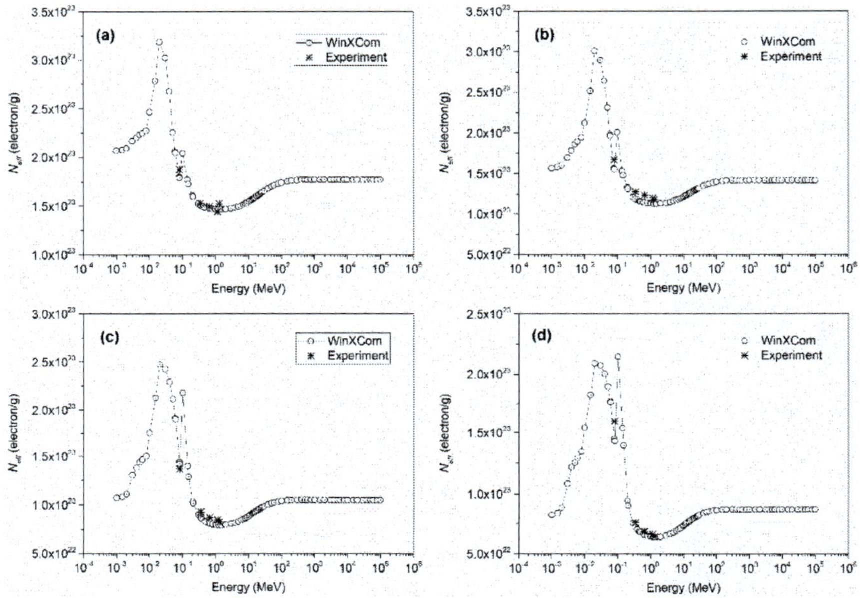

Figure 12. Variation of the experimental and WinXCom derived N_{eff} values of the (a) PP/5% Bi-2223, (b) PP/10% Bi-2223, (c) PP/20% Bi-2223, and (d) PP/30% Bi-2223 composites with increasing photon energy.

After these general comments on all Z_{eff} versus photon energy curves, the effects of Bi-2223 and YS additives on Zeff can be emphasized. If the related graphs are considered primarily in the context of the maximum value of Z_{eff}, it is seen that the lowest value of maximum Z_{eff} belongs to the pristine PP. Additionally, $(Z_{eff})_{max}$ = 5.33 for the pristine PP has been observed for the 0.002 MeV. With the increasing contribution percentage of both Bi-2223 and YS, the $(Z_{eff})_{max}$ value also tended to increase, and the highest $(Z_{eff})_{max}$ values have been observed in both composite families, with the highest additive contribution approximately at the same photon energy (0.02 MeV). ($(Z_{eff})_{max}$ = 15.196 for 30% Bi-2223 contribution and $(Z_{eff})_{max}$ = 9.848 for 30% YS additive.) Additionally, when the Z_{eff} values of the pristine PP and the composites are compared by choosing various spot photon energies, it has been determined that Z_{eff}

increased slowly and gradually with increasing Bi-2223 and YS additives. On the other hand, a remarkable point for the PP/Bi-2223 composites is also that the increasing Bi-2223 additive makes the K-edge discontinuity more obvious in the Z_{eff}-photon energy change graph. Moreover, since YS does not contain any elements having K-edge discontinuity such as tungsten or bismuth, any sharp increase in Z_{eff} has been observed in Region II for the YS reinforced PP composites.

Since N_{eff} is directly proportional to N_{eff} and inversely proportional to average atomic mass as the last parameter N_{eff} versus incoming photon energy curves have been drawn for the pristine PP, PP/Bi-2223, and PP/YS composites in Figures 11, 12, and 13, respectively.

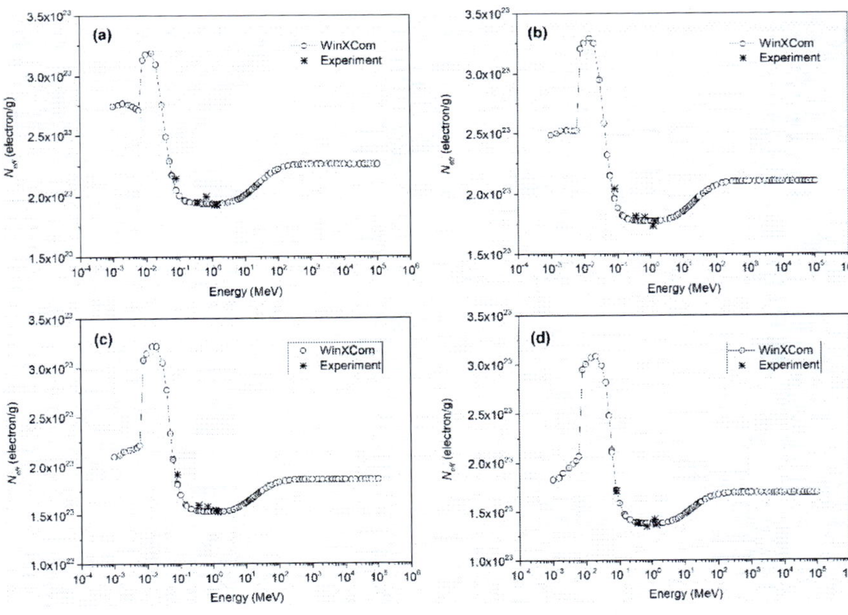

Figure 13. Variation of the experimental and WinXCom derived N_{eff} values of the (a) PP/5% YS, (b) PP/10% YS, (c) PP/20% YS, and (d) PP/30% YS composites with increasing photon energy.

Similar to the Z_{eff} analysis, N_{eff} versus photon energy graphs can be examined by using the highest N_{eff} value, since the high values of effective electron density can be considered as a good indicator of the increased possibility of photon-matter interactions. From this point of view, two concluding remarks can be expressed for the N_{eff}: The first is the highest effective electron density values have been observed extremely low photon energies and the N_{eff} is in the order of few 10^{23} electron/g. Secondly, it can be deduced that based on the highest N_{eff} values, it is more likely that the photon-matter interaction reaches its maximum level for 30% contribution of Bi-2223 and 10% contribution of YS.

CONCLUSION

The present study deals with the addition of ceramic (Bi-2223) and natural stone (YS) on the radiation shielding ability of commercial and low-cost PP polymer matrix for the first time. For this purpose, PP has been reinforced with Bi-2223 ceramic particles synthesized by routine solid-state reaction method and YS which was collected from Kayseri, Turkey. Then the PP/Bi-2223 and PP/YS composites having micro-sized reinforcing particles have been prepared by using a single screw extruder. The radiation shielding performances of the samples have been discussed experimentally and theoretically in the context of linear attenuation coefficient, half-value layer, mean free path, effective atomic number, and effective electron density. The experiments have been carried out by considering narrow beam geometry and using NaI(Tl) scintillation detector and the required equipment. The theoretical analysis has also been realized by running the WinXCom software. The study has been performed to reveal the radiation shielding capability of the novel composites against the photons having energies ranging from 81 keV to 1332 keV and emitted from Ba-133, Cs-137, and Co-60 point radioactive sources. Additionally, the theoretical study for determining the effective atomic number and the effective electron density has been made for many photon energies varying from 1×10^{-3} MeV to 1×10^{5} MeV.

As a result of all investigation, it has been determined that the two doping particles (Bi-2223 and YS) tested for the first time for PP significantly improve the radiation shielding property of PP, especially at low photon energies. Moreover, it has been observed that the polymers with the highest percentage of reinforcing particles produced in this study can either compete or exhibit better radiation shielding performance than most polymer composites and many polymer composites obtained after more demanding manufacturing conditions. In this respect, it is thought that this study will lead to more use of PP-based composites in radiation shielding research.

REFERENCES

AbuAlRoos, N. J., Amin, N. A. B., & Zainon, R. (2019). Conventional and new lead-free radiation shielding materials for radiation protection in nuclear medicine: A review. *Radiation Physics and Chemistry, 165*, 108439.

Afshar, M., Morshedian, J., & Ahmadi, S. (2019). Radiation attenuation capability and flow characteristics of HDPE composite loaded with W, MoS_2, and B_4C. *Polymer Composites, 40*(1), 149-158.

Akman, F., Ogul, H., Kaçal, M., Polat, H., Dilsiz, K., & Agar, O. (2021). Gamma attenuation characteristics of CdTe-Doped polyester composites. *Progress in Nuclear Energy, 131*, 103608.

Alavian, H., & Tavakoli-Anbaran, H. (2019). Study on gamma shielding polymer composites reinforced with different sizes and proportions of tungsten particles using MCNP code. *Progress in Nuclear Energy, 115*, 91-98.

Aldhuhaibat, M. J., Amana, M. S., Jubier, N. J., & Salim, A. (2021). Improved gamma radiation shielding traits of epoxy composites: Evaluation of mass attenuation coefficient, effective atomic and electron number. *Radiation Physics and Chemistry, 179*, 109183.

Ambika, M., Nagaiah, N., & Suman, S. (2017). *A new polyester based polymer composite for shielding soft gamma rays.* Paper presented at the AIP Conference Proceedings.

Avcıoğlu, S., Buldu, M., Kaya, F., Üstündağ, C. B., Kam, E., Menceloğlu, Y. Z., . . . Kaya, C. (2020). Processing and properties of boron carbide (B_4C) reinforced LDPE composites for radiation shielding. *Ceramics International, 46*(1), 343-352.

Azman, N. N., Siddiqui, S., Hart, R., & Low, I.-M. (2013). Effect of particle size, filler loadings and x-ray tube voltage on the transmitted x-ray transmission in tungsten oxide—epoxy composites. *Applied Radiation and Isotopes, 71*(1), 62-67.

Belgin, E. E., Aycik, G., Kalemtas, A., Pelit, A., Dilek, D., & Kavak, M. (2015). Preparation and characterization of a novel ionizing electromagnetic radiation shielding material: hematite filled polyester based composites. *Radiation Physics and Chemistry, 115*, 43-48.

Berger, M. (2010). *XCOM: photon cross sections database*. http://www.nist.gov/pml/data/xcom/index.cfm.

Berger, M. J., & Hubbell, J. (1987). *XCOM: Photon cross sections on a personal computer*. Retrieved from

Bichsel, H., & Schindler, H. (2020). The Interaction of Radiation with Matter. *Particle Physics Reference Library*, 5-44.

Cao, D., Yang, G., Bourham, M., & Moneghan, D. (2020). Gamma radiation shielding properties of poly (methyl methacrylate)/Bi_2O_3 composites. *Nuclear Engineering and Technology, 52*(11), 2613-2619.

Çağlar, M., Kayacık, H., Karabul, Y., Kılıç, M., Özdemir, Z. G., & İçelli, O. (2019). $Na_2Si_3O_7$/BaO composites for the gamma-ray shielding in medical applications: Experimental, MCNP5, and WinXCom studies. *Progress in Nuclear Energy, 117*, 103119.

El-Agawany, F., Mahmoud, K., Kavaz, E., El-Mallawany, R., & Rammah, Y. (2020). Evaluation of nuclear radiation shielding competence for ternary Ge–Sb–S chalcogenide glasses. *Applied Physics A, 126*(4), 1-11.

El-Khatib, A. M., Abbas, M. I., Abd Elzaher, M., Badawi, M. S., Alabsy, M. T., Alharshan, G. A., & Aloraini, D. A. (2019). Gamma attenuation coefficients of nano cadmium oxide/high density polyethylene composites. *Scientific reports, 9*(1), 1-11.

El-Sharkawy, R. M., Allam, E. A., El-Taher, A., Shaaban, E. R., & Mahmoud, M. E. (2021). Synergistic effect of nano-bentonite and nanocadmium oxide doping concentrations on assembly, characterization, and enhanced gamma-rays shielding properties of polypropylene ternary nanocomposites. *International Journal of Energy Research, 45*(6), 8942-8959.

Elgazzar, A. H., & Kazem, N. (2015). Biological effects of ionizing radiation. In *The pathophysiologic basis of nuclear medicine* (pp. 715-726): Springer.

Gerward, L., Guilbert, N., Jensen, K. B., & Levring, H. (2001). X-ray absorption in matter. Reengineering XCOM. *Radiation Physics and Chemistry, 60*(1-2), 23-24.

Gjorgieva, S., & Barandovski, L. (2016). *Measurement of the mass attenuation coefficient from 81 keV to 1333 keV for elemental materials Al, Cu and Pb.* Paper presented at the AIP Conference Proceedings.

Harish, V., Nagaiah, N., & Kumar, H. (2012). *Lead oxides filled isophthalic resin polymer composites for gamma radiation shielding applications.*

Hine, G. (1952). The effective atomic numbers of materials for various gamma ray processes. *Phys Rev, 85*, 725.

Issa, S. A., Zakaly, H. M., Pyshkina, M., Mostafa, M. Y., Rashad, M., & Soliman, T. (2021). Structure, optical, and radiation shielding properties of PVA–$BaTiO_3$ nanocomposite films: An experimental investigation. *Radiation Physics and Chemistry, 180*, 109281.

Karabul, Y., & İçelli, O. (2021). The Assessment of Usage of Epoxy Based Micro and Nano-Structured Composites Enriched with Bi_2O_3 and WO_3 Particles for Radiation Shielding. *Results in Physics*, 104423.

Kathem, H. O., Resen, M. J., & Jubier, N. J. (2018). Characterization of Radiation Attenuation Properties of 0.662 MeV Gamma Ray Energy for Epoxy Fe_3O_4 Composite Shields. *Indian Journal of Natural Sciences, 9*(50).

Korkut, T., Gencel, O., Kam, E., & Brostow, W. (2013). X-ray, gamma, and neutron radiation tests on epoxy-ferrochromium slag composites by experiments and Monte Carlo simulations. *International journal of polymer analysis and characterization, 18*(3), 224-231.

Körpınar, B., Öztürk, B. C., Çam, N. F., & Akat, H. (2020). Radiation shielding properties of Poly (hydroxylethyl methacrylate)/Tungsten (VI) oxide composites. *Materials Chemistry and Physics, 239,* 121986.

Kumar, A. P., Depan, D., Tomer, N. S., & Singh, R. P. (2009). Nanoscale particles for polymer degradation and stabilization—trends and future perspectives. *Progress in polymer science, 34*(6), 479-515.

La, L. B., Leong, Y. K., Leatherday, C., Au, P. I., Hayward, K. J., & Zhang, L.-C. (2016). X-ray protection, surface chemistry and rheology of ball-milled submicron Gd_2O_3 aqueous suspension. *Colloids and Surfaces A: Physicochemical and Engineering Aspects, 501*, 75-82.

Li, R., Gu, Y., Wang, Y., Yang, Z., Li, M., & Zhang, Z. (2017). Effect of particle size on gamma radiation shielding property of gadolinium oxide dispersed epoxy resin matrix composite. *Materials Research Express, 4*(3), 035035.

Li, R., Gu, Y., Zhang, G., Yang, Z., Li, M., & Zhang, Z. (2017). Radiation shielding property of structural polymer composite: continuous basalt fiber reinforced epoxy matrix composite containing erbium oxide. *Composites Science and Technology, 143*, 67-74.

Ma, K. C., Holmes, T. W., & Pourmorteza, A. (2019). K-edge imaging visualization of multi-material decomposition in CT using virtual mono-energetic images. *Paper presented at the 15th International Meeting on Fully Three-Dimensional Image Reconstruction in Radiology and Nuclear Medicine*.

Mahmoud, K., Lacomme, E., Sayyed, M., Özpolat, Ö., & Tashlykov, O. (2020). Investigation of the gamma ray shielding properties for polyvinyl chloride reinforced with chalcocite and hematite minerals. *Heliyon, 6*(3), e03560.

Manohara, S., Hanagodimath, S., Thind, K., & Gerward, L. (2008). On the effective atomic number and electron density: a comprehensive set of formulas for all types of materials and energies above 1 keV. *Nuclear Instruments and Methods in Physics Research Section B: Beam Interactions with Materials and Atoms, 266*(18), 3906-3912.

Mirzaei, M., Zarrebini, M., Shirani, A., Shanbeh, M., & Borhani, S. (2019). X-ray shielding behavior of garment woven with melt-spun polypropylene monofilament. *Powder Technology, 345*, 15-25.

Mori, H., Oku, Y., Mannami, Y., & Kunisada, T. (2020). Iron (III) Oxide-Based Ceramic Material for Radiation Shielding. *Ceramics, 3*(2), 258-264.

Muthamma, M., Gudennavar, B. S., & Gudennavar, S. B. (2020). Attenuation parameters of polyvinyl alcohol-tungsten oxide composites at the photon energies 5.895, 6.490, 59.54 and 662 keV. *Polish Journal of Medical Physics and Engineering, 26*(2), 77-85.

Nambiar, S., Osei, E. K., & Yeow, J. T. (2013). Polymer nanocomposite-based shielding against diagnostic X-rays. *Journal of applied polymer science, 127*(6), 4939-4946.

Oto, B., Yıldız, N., Akdemir, F., & Kavaz, E. (2015). Investigation of gamma radiation shielding properties of various ores. *Progress in Nuclear Energy, 85*, 391-403.

Parlar, Z., Abdlhamed, A., & Akkurt, İ. (2019). Gamma-ray-shielding properties of composite materials made of recycled sport footwear. *International Journal of Environmental Science and Technology, 16*(9), 5113-5116.

Petrucci, R. H. (2017). *General chemistry: principles and modern applications*: Pearson.

Szajerski, P., Zaborski, M., Bem, H., Baryn, W., & Kusiak, E. (2014). Optimization of the heavy metal (Bi–W–Gd–Sb) concentrations in the elastomeric shields for computer tomography (CT). *Journal of radioanalytical and nuclear chemistry, 300*(1), 385-391.

Şahin, N., Bozkurt, M., Karabul, Y., Kılıç, M., & Özdemir, Z. G. (2021). Low cost radiation shielding material for low energy radiation applications: Epoxy/Yahyali Stone composites. *Progress in Nuclear Energy, 135*, 103703.

Walker, J. S., & Wellock, T. R. (2010). A Short History of Nuclear Regulation. *US Nuclear Regulatory Commission, Washington, DC*, 1946-2009.

Biographical Sketches

Mehmet Kiliç

Affiliation: Yildiz Technical University.

Education: PhD.

Research and Professional Experience: Assoc. Prof. Dr. Mehmet Kılıç has been working as a scientist at Yildiz Technical University, Department of Physics since 2009. He has obtained his PhD and MSc degrees from Yildiz Technical University. He has published 30 scientific papers in his academic carrier. His main areas of interest are photonics, dielectric spectroscopy, and mechanical and radiation shielding properties of polymer composites both in micro-structure and nano-structure.

Publications from the Last 3 Years:

1. Kılıç, M., Güven Özdemir, Z., Alkan, Ü., Karabul, Y., and İçelli, O., "The Effects of Gamma Irradiation Strength and Basalt Additive Content on Mechanical Performance and Dielectric Response of Polypropylene Films," *Journal of Applied Polymer Science*, 136 (18), 47414 (1-14) (2019).

2. Kılıç, M., Karabul, Y., Güven Özdemir, Z., Süngü Mısırlıoğlu, B., and İçelli, O., "Improved Dielectric and Electrical Properties of PANI Achieved By Using Low Cost Mineral Additive," *IEEE Transactions on Dielectrics and Electrical Insulation*, 26(1), 300-307 (2019).
3. Kılıç. M., Çağlar, M., Karabul, Y., Güven Özdemir, Z., and İçelli, O., "Dielectric and Gamma Shielding Properties of Sodium Silicate Glassy Structure Modified with Tungsten," *Radiation Effects and Defects in Solids*, 174 (3–4), 229–247 (2019).
4. Demirbay, T., Çağlar, M., Karabul, Y., Kılıç, M., İçelli, O., and Güven Özdemir, Z., "Availability of Water Glass/Bi_2O_3 Composites in Dielectric and Gamma-Ray Screening Applications" *Radiation Effects and Defects in Solids*, 174 (5–6), 419–434 (2019).
5. Aslan Çataltepe, Ö., Güven Özdemir, Z., Kılıç, M., and Onbaşlı, Ü., "Impact of oxygen doping on AC Impedance, Complex Dielectric and Electrical Modulus Spectra of a Hg-Ba-Ca-Cu-O Ceramic," *Materials Testing* 61 (7), 685-689 (2019).
6. Çağlar, M., Kayacık, H., Karabul, Y., Kılıç, M., Güven Özdemir, Z., and İçelli, O., $Na_2Si_3O_7$/BaO Composites for the Gamma-Ray Shielding in Medical Applications: Experimental, MCNP5, and WinXCom Studies, *Progress in Nuclear Energy*, 117, 103119 (2019).
7. Bulgurcuoğlu, A.E., Karabul, Y., Kılıç, M., Güven Özdemir, Z, Erdönmez, S., Süngü Mısırlıoğlu, B., Okutan, M. and İçelli, O., "Structural Analysis and Dielectric Relaxation Mechanism of Conducting Polymer/ Volcanic Basalt Rock Composites," *Materials Science-Poland*, 37(3), 353-363 (2019).
8. Güven Özdemir, Z., Kılıç, M., Karabul, Y., Erdönmez, S., and İçelli, O., "The Influence of the Partially Europium Substitution on the AC Electrical Parameters of $BiSr_2CaCu_2O_{6.5}$," *Processing and Applications of Ceramics*, 13(4), 323-332 (2019).
9. Kılıç M., "Natural additive material for desirable dielectric properties of polypyrrole: Limestone," *Synthetic Metals*, 260, 116297 (2020).

10. Güven Özdemir, Z., Kılıç, M., Karabul, Y., Süngü Mısırlıoğlu, B., Çakır, Ö and Kahya, N. D, "A Transition in the Electrical Conduction Mechanism of $CuO/CuFe_2O_4$ Nanocomposites," *Journal of Electroceramics*, 44(1-2), 1-15 (2020).
11. Kılıç, M., Eyecioğlu, Ö., Güven Özdemir, Z., and Alkan Ü., "Estimation of Dielectric Parameters of LDPE/PANI Composite Films Depending on Temperature and PANI Additive Concentration by GRNN," *Journal of the Faculty of Engineering and Architecture of Gazi University*, 35(2), 1077-1088 (2020).
12. Kılıç, M., Güven Özdemir, Z., Karabul, Y., Karataş, Ö., and Aslan Çataltepe, Ö., "Negative Real Permittivity in $(Bi_{0.3}Eu_{0.7})Sr_2CaCu_2O_{6.5}$ Ceramic," *Physica B-Condensed Matter*, 584, 412080 (2020).
13. Alkan, Ü., Kılıç, M., Karabul, Y., and Güven Özdemir, Z., "Mechanical and Dielectric Behavior of LDPE/Bi-1212 Films," *Materials Testing* 62 (9), 913-920 (2020).
14. Kılıç M., Kahya, N. D., Süngü Mısırlıoğlu, B., Çakır, Ö., and Güven Özdemir, Z., "Dielectric and Magnetic Properties of $CuFe_2O_4/CuO$ Nanocomposites," *Ferroelectrics*, 571 (1), 183-199 (2021).
15. Erdönmez, S., Karabul, Y., Kılıç, M., Güven Özdemir, Z., and Esmer, K., "Structural Characterization and Dielectric Parameters of Polyindole/WO_3 Nanocomposites," *Polymer Composites*, 42 (3), 1347-1355 (2021).
16. Şahin, N., Bozkurt, M., Karabul, Y., Kılıç, M., and Güven Özdemir, Z., "Low Cost Radiation Shielding Material for Low Energy Radiation Applications: Epoxy/Yahyali Stone Composites," *Progress in Nuclear Energy*, 135, 103703 (2021).
17. Çağlar, M., Karabul, Y., Kılıç, M., Güven Özdemir, Z., İçelli, O., "$Na_2Si_3O_7$/Ag Micro and Nano-Structured Glassy Composites: The Experimental and MCNP Simulation Surveys of Their Radiation Shielding Performances," *Progress in Nuclear Energy*, 139, 103855 (2021).

Yaşar Karabul

Affiliation: Yildiz Technical University.

Education: PhD.

Research and Professional Experience: Dr. Yaşar KARABUL has been working as a scientist at Yildiz Technical University, Department of Physics since 2014. He has obtained his PhD and MSc degrees from Yildiz Technical University and İstanbul University, respectively. He has published 33 scientific papers in his academic carrier. His main areas of interest are nuclear physics, dielectric spectroscopy, and nanomaterials.

Publications from the Last 3 Years:

1. Kılıç, M., Güven Özdemir, Z., Alkan, Ü., Karabul, Y., and İçelli, O., "The Effects of Gamma Irradiation Strength and Basalt Additive Content on Mechanical Performance and Dielectric Response of Polypropylene Films," *Journal of Applied Polymer Science*, 136 (18), 47414 (1-14) (2019).
2. Kılıç, M., Karabul, Y., Güven Özdemir, Z., Süngü Mısırlıoğlu, B., and İçelli, O., "Improved Dielectric and Electrical Properties of PANI Achieved By Using Low Cost Mineral Additive," *IEEE Transactions on Dielectrics and Electrical Insulation*, 26(1), 300-307 (2019).
3. Kılıç. M., Çağlar, M., Karabul, Y., Güven Özdemir, Z., and İçelli, O., "Dielectric and Gamma Shielding Properties of Sodium Silicate Glassy Structure Modified with Tungsten," *Radiation Effects and Defects in Solids*, 174 (3–4), 229–247 (2019).
4. Demirbay, T., Çağlar, M., Karabul, Y., Kılıç, M., İçelli, O., and Güven Özdemir, Z., "Availability of Water Glass/Bi_2O_3 Composites in Dielectric and Gamma-Ray Screening Applications" *Radiation Effects and Defects in Solids*, 174 (5–6), 419–434 (2019).

5. Çağlar, M., Kayacık, H., Karabul, Y., Kılıç, M., Güven Özdemir, Z., and İçelli, O., $Na_2Si_3O_7$/BaO Composites for the Gamma-Ray Shielding in Medical Applications: Experimental, MCNP5, and WinXCom Studies, *Progress in Nuclear Energy*, 117, 103119 (2019).
6. Bulgurcuoğlu, A. E., Karabul, Y., Kılıç, M., Güven Özdemir, Z, Erdönmez, S., Süngü Mısırlıoğlu, B., Okutan, M. and İçelli, O., "Structural Analysis and Dielectric Relaxation Mechanism of Conducting Polymer/ Volcanic Basalt Rock Composites," *Materials Science-Poland*, 37(3), 353-363 (2019).
7. Güven Özdemir, Z., Kılıç, M., Karabul, Y., Erdönmez, S., and İçelli, O., "The Influence of the Partially Europium Substitution on the AC Electrical Parameters of $BiSr_2CaCu_2O_{6.5}$," *Processing and Applications of Ceramics*, 13(4), 323-332 (2019).
8. Eyecioglu O., El-Khayatt A. M., Karabul Y., Cağlar M., Toker O., İçelli O., "BXCOM: a software for computation of radiation sensing," *Radiation Effects and Defects in Solids*, 174, 506-518, (2019).
9. Toker O., Topdağı Ö., Bakırdere S., Bursalıoglu E. O., Öz E., Eyecloglu O., Karabul, Y., Çaglar, M. and Içelli, O., "Determination of Se, Cr, Mn, Zn, Co, Na, and K in Blood Samples of Breast Cancer Patients to Investigate Their Variation Using ICP-MS and ICP-OES," *Atomic Spectroscopy*, 40, 11-16 (2019).
10. Güven Özdemir, Z., Kılıç, M., Karabul, Y., Süngü Mısırlıoğlu, B., Çakır, Ö and Kahya, N. D, "A Transition in the Electrical Conduction Mechanism of CuO/$CuFe_2O_4$ Nanocomposites," *Journal of Electroceramics*, 44(1-2), 1-15 (2020).
11. Kılıç, M., Güven Özdemir, Z., Karabul, Y., Karataş, Ö., and Aslan Çataltepe, Ö., "Negative Real Permittivity in $(Bi_{0.3}Eu_{0.7})Sr_2CaCu_2O_{6.5}$ Ceramic," *Physica B-Condensed Matter*, 584, 412080 (2020).
12. Alkan, Ü., Kılıç, M., Karabul, Y., and Güven Özdemir, Z., "Mechanical and Dielectric Behavior of LDPE/Bi-1212 Films," *Materials Testing* 62 (9), 913-920 (2020).

13. Erdönmez, S., Karabul, Y., Kılıç, M., Güven Özdemir, Z., and Esmer, K., "Structural Characterization and Dielectric Parameters of Polyindole/WO_3 Nanocomposites," *Polymer Composites*, 42 (3), 1347-1355 (2021).
14. Şahin, N., Bozkurt, M., Karabul, Y., Kılıç, M., and Güven Özdemir, Z., "Low Cost Radiation Shielding Material for Low Energy Radiation Applications: Epoxy/Yahyali Stone Composites," *Progress in Nuclear Energy*, 135, 103703 (2021).
15. Karabul Y. and İçelli O., "The assessment of usage of epoxy based micro and nano-structured composites enriched with Bi_2O_3 and WO_3 particles for radiation shielding," *Results in Physics*, 26, 104423 (2021).
16. Çağlar, M., Karabul, Y., Kılıç, M., Güven Özdemir, Z., İçelli, O., "$Na_2Si_3O_7$/Ag Micro and Nano-Structured Glassy Composites: The Experimental and MCNP Simulation Surveys of Their Radiation Shielding Performances," *Progress in Nuclear Energy*, 139, 103855 (2021).

Zeynep Güven Özdemir

Affiliation: Yildiz Technical University.

Education: PhD.

Research and Professional Experience: Assoc. Prof. Dr. Zeynep Güven Özdemir has been working as a scientist at Yildiz Technical University, Department of Physics for over 20 years. She has obtained her Ph.D. and MSc degrees from Marmara University and Yildiz Technical University, respectively. She has published more than 50 scientific papers and 4 book chapters in her academic carrier. Her main areas of interest are high temperature superconductivity, electrical properties of liquid crystals, dielectric and radiation shielding properties of polymer nanocomposites.

Publications from the Last 3 Years:

1. Kılıç, M., Güven Özdemir, Z., Alkan, Ü., Karabul, Y., and İçelli, O., "The Effects of Gamma Irradiation Strength and Basalt Additive Content on Mechanical Performance and Dielectric Response of Polypropylene Films," *Journal of Applied Polymer Science*, 136 (18), 47414 (1-14) (2019).
2. Kılıç, M., Karabul, Y., Güven Özdemir, Z., Süngü Mısırlıoğlu, B., and İçelli, O., "Improved Dielectric and Electrical Properties of PANI Achieved By Using Low Cost Mineral Additive," *IEEE Transactions on Dielectrics and Electrical Insulation*, 26(1), 300-307 (2019).
3. Kılıç. M., Çağlar, M., Karabul, Y., Güven Özdemir, Z., and İçelli, O., "Dielectric and Gamma Shielding Properties of Sodium Silicate Glassy Structure Modified with Tungsten," *Radiation Effects and Defects in Solids*, 174 (3–4), 229–247 (2019).
4. Demirbay, T., Çağlar, M., Karabul, Y., Kılıç, M., İçelli, O., and Güven Özdemir, Z., "Availability of Water Glass/Bi_2O_3 Composites in Dielectric and Gamma-Ray Screening Applications" *Radiation Effects and Defects in Solids*, 174 (5–6), 419–434 (2019).
5. Aslan Çataltepe, Ö., Güven Özdemir, Z., Kılıç, M., and Onbaşlı, Ü., "Impact of oxygen doping on AC Impedance, Complex Dielectric and Electrical Modulus Spectra of a Hg-Ba-Ca-Cu-O Ceramic," *Materials Testing* 61 (7), 685-689 (2019).
6. Çağlar, M., Kayacık, H., Karabul, Y., Kılıç, M., Güven Özdemir, Z., and İçelli, O., $Na_2Si_3O_7$/BaO Composites for the Gamma-Ray Shielding in Medical Applications: Experimental, MCNP5, and WinXCom Studies, *Progress in Nuclear Energy*, 117, 103119 (2019).
7. Bulgurcuoğlu, A.E., Karabul, Y., Kılıç, M., Güven Özdemir, Z, Erdönmez, S., Süngü Mısırlıoğlu, B., Okutan, M. and İçelli, O., "Structural Analysis and Dielectric Relaxation Mechanism of Conducting Polymer/ Volcanic Basalt Rock Composites," *Materials Science-Poland*, 37(3), 353-363 (2019).

8. Güven Özdemir, Z., Kılıç, M., Karabul, Y., Erdönmez, S., and İçelli, O., "The Influence of the Partially Europium Substitution on the AC Electrical Parameters of $BiSr_2CaCu_2O_{6.5}$," *Processing and Applications of Ceramics*, 13(4), 323-332 (2019).
9. Erdönmez S. and Güven Özdemir, Z. "The Mechanical and Dielectric Properties of Polypropylene Reinforced with $BiSr_2CaCu_2O_{6.5}$ Ceramic," *Polymer Bulletin* 77, 1793–1812 (2020).
10. Güven Özdemir, Z., Kılıç, M., Karabul, Y., Süngü Mısırlıoğlu, B., Çakır, Ö and Kahya, N.D, "A Transition in the Electrical Conduction Mechanism of $CuO/CuFe_2O_4$ Nanocomposites," *Journal of Electroceramics*, 44(1-2), 1-15 (2020).
11. Kılıç, M., Eyecioğlu, Ö., Güven Özdemir, Z., and Alkan Ü., "Estimation of Dielectric Parameters of LDPE/PANI Composite Films Depending on Temperature and PANI Additive Concentration by GRNN," *Journal of the Faculty of Engineering and Architecture of Gazi University*, 35(2), 1077-1088 (2020).
12. Güven Özdemir, Z., Şakar Daşdan, D., Kavak, P., Pionteck, J., Potschke, P., Voit, B., and Süngü Mısırlıoğlu, B. "MWCNT Induced Negative Real Permittivity in a Copolyester of Bisphenol-A with Terephthalic and Isophthalic Acids," *Materials Research Express*, 7, 015337 (2020).
13. Kılıç, M., Güven Özdemir, Z., Karabul, Y., Karataş, Ö., and Aslan Çataltepe, Ö., "Negative Real Permittivity in $(Bi_{0.3}Eu_{0.7})Sr_2CaCu_2O_{6.5}$ Ceramic," *Physica B-Condensed Matter*, 584, 412080 (2020).
14. Alkan, Ü., Kılıç, M, Karabul, Y., and Güven Özdemir, Z., "Mechanical and Dielectric Behavior of LDPE/Bi-1212 Films," *Materials Testing* 62 (9), 913-920 (2020).
15. Kılıç M., Kahya, N. D., Süngü Mısırlıoğlu, B., Çakır, Ö., and Güven Özdemir, Z., "Dielectric and Magnetic Properties of $CuFe_2O_4/CuO$ Nanocomposites," *Ferroelectrics*, 571 (1), 183-199 (2021).
16. Erdönmez, S., Karabul, Y., Kılıç, M., Güven Özdemir, Z., and Esmer, K., "Structural Characterization and Dielectric Parameters

of Polyindole/WO_3 Nanocomposites," *Polymer Composites*, 42 (3), 1347-1355 (2021).

17. Şahin, N., Bozkurt, M., Karabul, Y., Kılıç, M., and Güven Özdemir, Z., "Low Cost Radiation Shielding Material for Low Energy Radiation Applications: Epoxy/Yahyali Stone Composites," *Progress in Nuclear Energy*, 135, 103703 (2021).
18. Çağlar, M., Karabul, Y., Kılıç, M., Güven Özdemir, Z., İçelli, O., "$Na_2Si_3O_7$/Ag Micro and Nano-Structured Glassy Composites: The Experimental and MCNP Simulation Surveys of Their Radiation Shielding Performances," *Progress in Nuclear Energy*, 139, 103855 (2021).

In: Polypropylene
Editor: Théodore Marleau
ISBN: 978-1-68507-378-7

Chapter 5

FRACTURE IN POLYPROPYLENE NANOCOMPOSITE

M. R. Nakhaei[1,*] and G. Naderi[2]
[1]Mechanics and Energy Faculty,
Shahid Beheshti University, Tehran, Iran
[2]Iran Polymer and Petrochemical Institute, Tehran, Iran

ABSTRACT

In recent years, one of the advances in plastic and composite industries has been the development and commodification of polymer-based nanocomposites. Polymer nanocomposites have recently become the subject of intensive research due to their excellent properties such as electrical and thermal conductivity, toughness and stiffness, mechanical strength, higher barrier to moisture and gases, and superior flame retardancy when compared with the pure polymer or conventional composites (micro-and macro composite). Mechanical properties of the composite and nanocomposite polymers, including compressive modulus, rate-dependent modulus, stress relaxation behaviors, dynamic viscoelastic properties, and fracture mechanics, strongly depend on the

* Corresponding Author's E-mail: m_nakhaei@sbu.ac.ir.

homogeneity of the filler dispersion. In this chapter, we will provide a brief overview of polypropylene nanocomposites, types of nanofiller, and their applications in the industry. The basic principle of fracture mechanics and crack propagation in polymer materials is described. The effect of different nanoparticles on mechanical properties, morphology, and fracture mechanism of polypropylene nanocomposites are discussed.

Keywords: polypropelene, nanocomposite, fracture mechanism, crack propagation, nanofiller

INTRODUCTION

The interest in polymer nanocomposites has been increasing since Toyota central research laboratories illustrated the advantages and commercial applications of nylon 6/clay nanocomposites in 1987 [1, 2]. In recent years, among the polymer nanocomposites, those based on polypropylene (PP) have attracted considerable interest because PP is a thermoplastic that has found wide applications due to its low price, low density, high thermal stability, high performance, and easy processin [3, 4]. To improve its mechanical and thermal properties, nanoparticles such as nanoclay, single (SWNT) and multi-walled (MWNT) carbon nanotube, Titanium oxide (Tio_2), and Graphene particles at very low loading levels have been used as fillers in polypropylene [5, 6]. The property changes in polymer nanocomposites compared to the base polymer are generally enhanced mechanical properties such as stiffness, tensile strength, modulus strength, and toughness [7]. Also, the friction coefficient is reduced and the abrasion resistance increases. Furthermore, the heat resistance of the material is generally much improved, with the mold shrinkage, thermal expansion coefficient, and dimensional change rate each decreased, leading to improved dimensional accuracy and stability. Functional properties, including gas-barrier properties and flame retardancy, are also improved [6, 8].

Polypropylene Nanocomposite

Polypropylene (PP) nanocomposites are described as a mixture of two or more materials, where the matrix is a PP, and the nanofillers have sizes (at least in one direction) of nanometer order (between 1 to 100 nm) which are dispersed homogeneously in PP [9]. The addition of a very small proportion of nanoparticles such as nanoclay, nanocellulose, halloysite nanotube (HNT), Carbon nanostructure (carbon nanotube, forlorn, and graphene), and titanium oxide (Tio_2) particles (Maximum 5 wt %) improves mechanical, barrier, flammability and thermal properties without affecting their processability [10]. The reinforcing effect of nanoparticles is related to several factors, such as nature and type of nanoparticle, particle size, particle aspect ratio, particle distribution, particle orientation, properties of the polymer matrix, and concentration of polymer and nanoparticles. One of the main reasons for using nanoparticles is the large surface-to-volume ratio that increases the number of particle-matrix interactions, thus increasing the effects on the overall material properties. The most significant challenge of using different nanofiller in the polymer matrix is ensuring their sufficient dispersion and uniform distribution in the base material [11]. According to Figure 1, nanofillers are classified into three categories, including particles having three dimensions in nanoscale as opposed to fibers/tubes and platelets having two- and one-dimension in nano-scale, respectively.

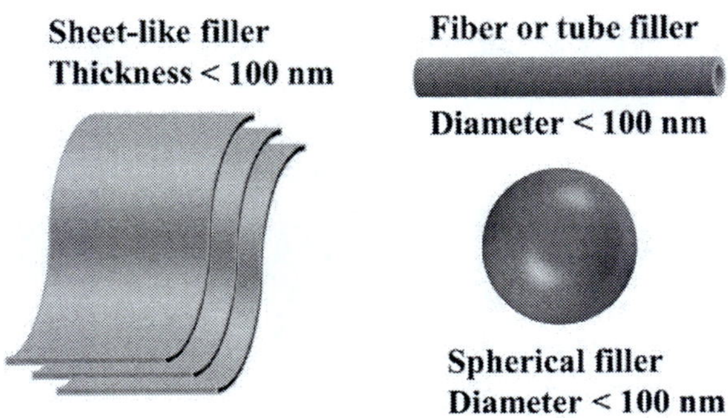

Figure 1. Schematic diagrams of three types of nano-scale fillers.

MANUFACTURING PROCESSES

Polymer nanocomposites can be produced by three methods: solution blending, in situ polymerization, and melt blending. An appropriate method is selected according to the nanofiller, polymeric matrix, and desired properties for the final products [12, 13].

Solution Blending

Solution blending involves dissolving a polymer in a solvent and suspending the filler in the same or another compatible solvent. Various solvents used include acetone, tetrahydrofuran, toluene, dimethylformamide, chloroform, and cyclohexane, etc. During the solution-blending process, high-speed shear mixing, ultrasonication, or stirring is usually employed to mix the filler suspension with the polymer solution [14]. There are some problems for the solution blending from the economic and environmental point of view. A proper decision must be taken to choose the correct method according to the situation and the desired product. The main limitations are aggregation and environmental constraints. Some of the benefits of using solution blending are easy operation, reduction in the permeability of gases, and general technique for all types of nanofillers and to both thermoset and thermoplastic polymers [15]. Figure 2 shows a schematic of the solution processing method for polymer/ CNT nanocomposite film production.

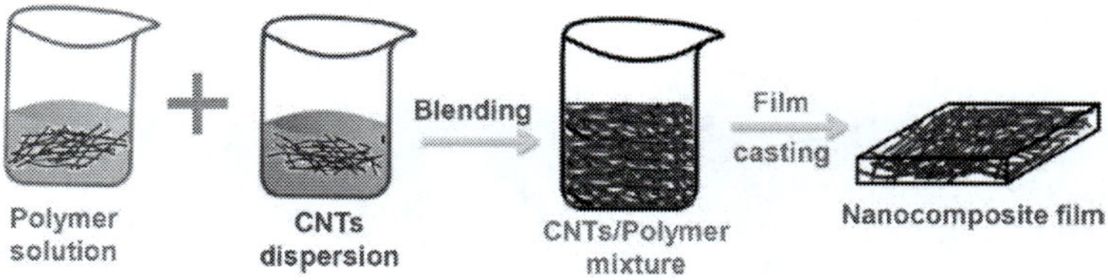

Figure 2. Schematic of a solution processing method for nanocomposite film production.

In Situ Polymerization

The mechanism of in situ polymerization is to blend the nanofiller with the monomer or prepolymer to form a stable chemical bond and then add a suitable initiator to initiate polymerization under heat or radiation conditions. It ensures uniform dispersion of filler particles in the polymer matrix. It makes the interaction between filler and polymer in the prepared composite material more signicant conducive to stress transfer [16, 17]. Schematic of in situ polymerization method illustrated in Figure 3. The advantages of this method are 1) the one-step synthesis of the metallocene polymer nanocomposites, 2) improved compatibility of the nanofiller and the polymer matrix, and 3) enhanced nanofiller dispersity. On the other hand, this approach is not generally applicable to all the polymer types, and polymerization is often incomplete [18].

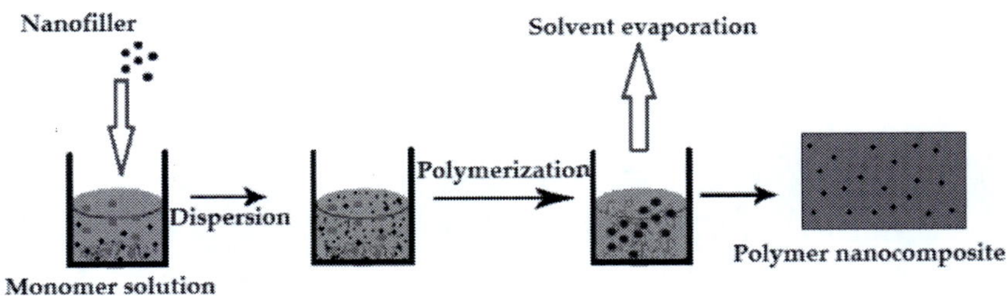

Figure 3. Schematic of in situ polymerization method.

Melt Blending

Melt blending is a more adaptable technique for thermoplastic and elastomeric polymeric matrix. This method is found to be environmentally friendly, cost-effective, and best for mass production [19]. The good thing in this method is that it does not require any type of solvent, and the nanofillers can be directly mixed in the molten polymer matrix. Dispersing one polymer in another immiscible polymer can best be achieved by intense mechanical stirring in compounders at elevated temperatures when all components are in a molten state [20]. The conventional methods are used to mix of the thermoplastic polymer with the nanofillers such as clay, HNT, graphene, CNT, and SIC. Examples of such methods are extrusion and internal mixer (Figures 4 and 5).

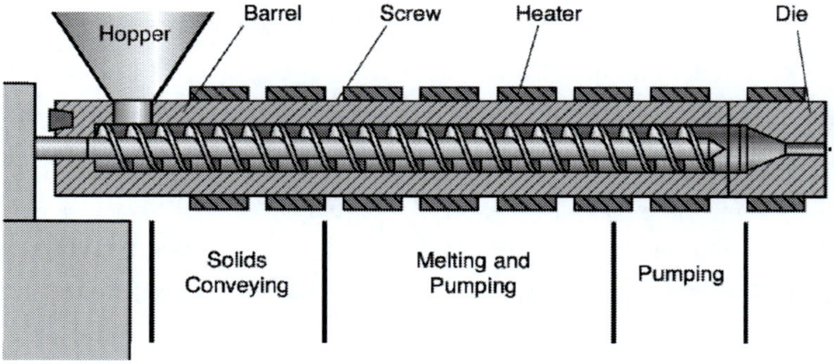

Figure 4. Schematic representation of a single-screw hot melt extruder.

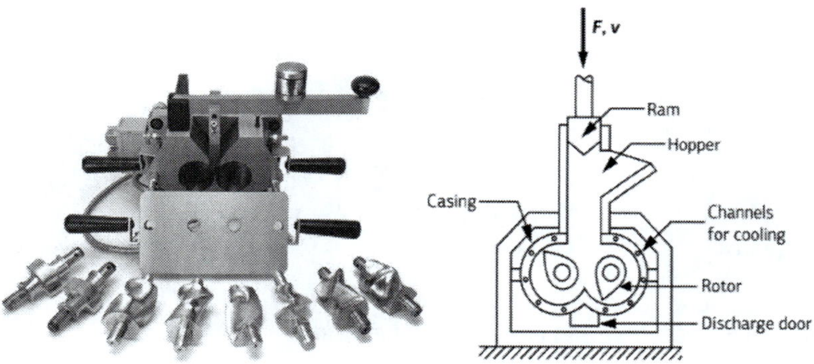

Figure 5. Schematic representation of an internal mixer for polymer blending.

POLYPROPYLENE NANOCOMPOSITE WITH DIFFERENT NANOFILLER

Nanoclay

Clay minerals are a class of phyllosilicates that usually form because of the chemical weathering of other silicate minerals at the earth's surface. According to the characteristics of layer type (1:1, or 2:1), the magnitude of net layer charge (x) per formula unit, and the type of interlayer species, the clay minerals can be classified into seven groups:

1. kaolin-serpentine,
2. pyrophyllite-talc,
3. smectite,
4. vermiculite,
5. mica,
6. chlorite,
7. Interstratified clay minerals [21, 22].

Clay has been used as a filler in polymers for many years, such as kaolin and bentonite; however, it plays a little functional role and mainly serves economic considerations. Nanoclays modify the properties of polymers such as nylon 6, EVA, epoxy, PET, PE, and PP. Improvement in clarity, stiffness, thermal stability, barrier properties (moisture, solvents, vapors, gases, and flavors), chemical properties, flame resistance, scratch resistance, and dimensional stability are noted. It was not until the early 1990s that clay received great attention from the polymer industry when Toyota's researchers discovered the possibility of constructing a nanostructure with a polymer and an organoclay [23].

In general, polymer-clay composites may be divided into the following three categories:

1. Conventional composite: where the clay particles exist in their original aggregated states, without intercalation of polymer matrix into clay galleries.
2. Intercalated nanocomposite: where polymers are inserted into the galleries of clay;
3. Exfoliated (delaminated) nanocomposite: where individual 1 nm-thick layers of clay are dispersed in a continuous polymer matrix [24, 25].

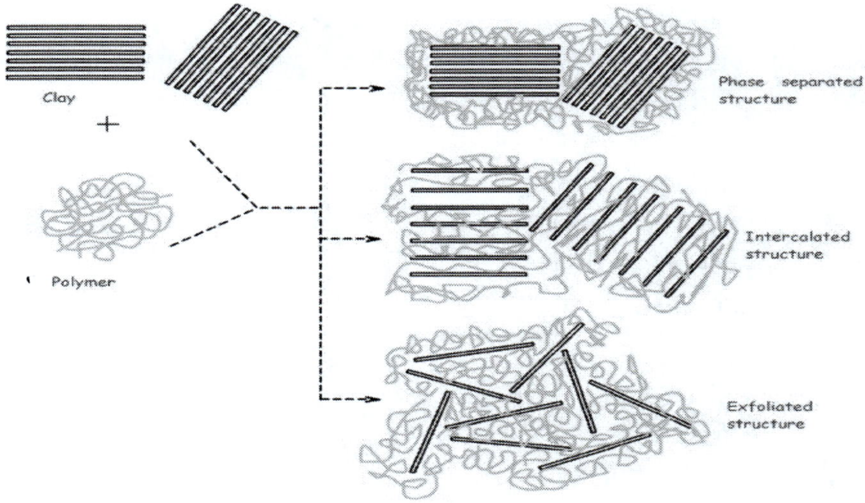

Figure 6. Various structures of polymer/clay composites.

The exfoliated configuration is of particular interest because it maximizes the interactions between polymers and clay. Thus, this configuration should result in maximum changes in mechanical and physical properties.

Techniques of Characterization

To investigate the morphology and properties of polymer nanocomposite characterization is necessary. Characterization is generally done for the following purposes:

- To understand the effect of filler surface correction on composite properties and filler dispersion.
- To analyze the quality of dispersion and orientation/alignment of fillers in the polymer matrix.
- To study the filler grafting interaction with the polymer chains and the chemical reactions between the two.

- In addition, to study the broad array of properties to ascertain the ability of nanocomposites for the application purpose [26, 27].

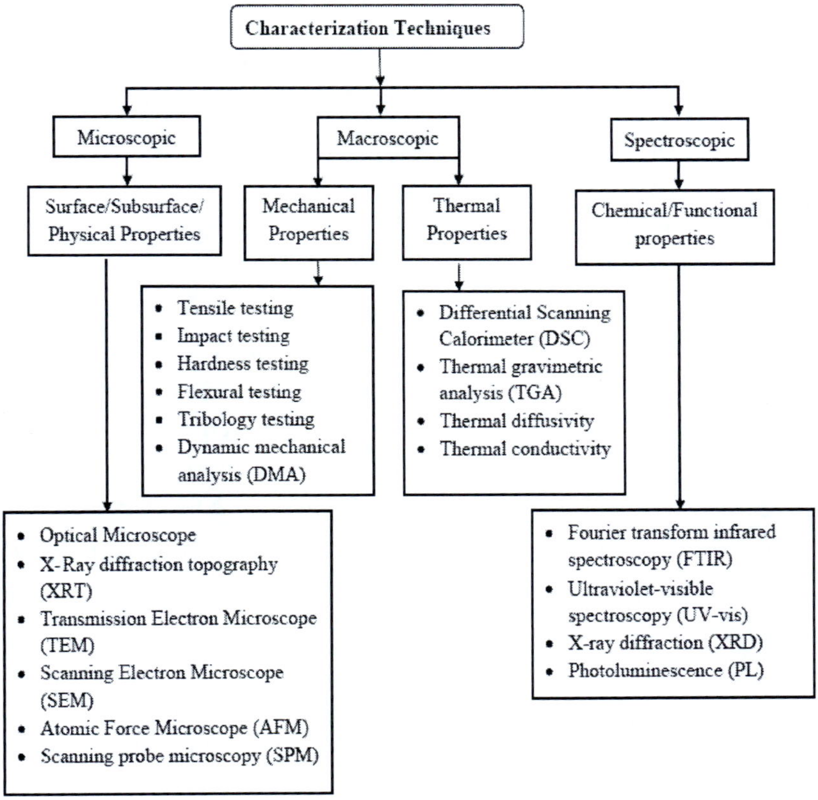

Figure 7. Various techniques of characterization.

According to Figure 7, there are generally three techniques of characterization of materials microscopic, macroscopic, and spectroscopic. Transmission electron microscope (TEM), scanning electron microscope (SEM), and optical microscope (OM) is the most commonly microscopic techniques for the investigation of the surface and sub-surface structure. Macroscopic techniques characterize various macroscopic properties of materials such as toughness, fatigue, creep, hardness, rheological, thermal, and tribological, etc. Mainly, Thermogravimetric analysis (TGA) and differential scanning calorimetry

(DSC) techniques are used for thermal characterization. Chemical and functional properties such as chemical composition, photoelectric properties, composition variation, and crystal structure of composites are examined using spectroscopic techniques. X-ray diffraction (XRD), Fourier transform infrared spectroscopy (FTIR), and X-ray photoelectron spectroscopy (XPS) are a few commonly used techniques in functional and chemical characterization [28, 29].

Fracture Behavior

Fracture toughness, which reflects the material's fracture resistance, is an important material property for engineering applications. To determine the fracture toughness, many fracture analysis methods were developed, such as K and J-Integral. Those developed or developing fracture mechanics can provide a reproducible way to characterize the fracture property, which can guide the selection of suitable materials and prevent the events happening due to the fracture or failure. Generally speaking, there are three main categories, i.e.,:

- Linear elastic fracture mechanics (LEFM),
- Elastic-plastic fracture mechanics (EPFM),
- Post-yield fracture mechanics (PYFM) [30, 31].

Linear Elastic Fracture Mechanics (LEFM)

Linear Elastic Fracture Mechanics was first used for metallic materials, nowadays; it has also been widely accepted to characterize the fracture toughness for polymeric materials and composites. According to LEFM theory, the fracture toughness can be represented by the stress intensity factor K or strain-energy release rate G. The method is only applicable to brittle materials with a very small yielding area ahead the crack tip, i.e., the inelastic deformation must be confined to a very small region at the crack tip [32].

The stress intensity factor K is based on the stress concentrations around the crack tip. When K reaches a critical value, failure will occur, and the critical stress intensity factor will be:

$$K_c = \sigma_y^2 a_c Y^2 \tag{1}$$

where σ_y is material yield stress, a_c is the critical flaw size, and Y is a sample geometry factor, experimental procedures for the determination of K_c are standardized by ASTM E399-83, and a modified version (D5045) for polymeric materials is available. This standard defines an apparent crack initiation load, P_Q, by 5% secant construction in the load-displacement curve, which corresponds to crack growth through around 2% ligament length in the absence of plastic deformation. Provisional fracture toughness K_Q can be calculated by the following formula once the provisional critical load P_Q and the crack length a are determined.

$$K_Q = \frac{P_Q}{B\sqrt{W}} f\left(\frac{a}{W}\right) \tag{2}$$

where f $\left(\frac{a}{W}\right)$ is a function of $\frac{a}{W}$ and B is a shape factor. When the following requirements can be satisfied, the K_Q can be treated as K_{IC}, the critical stress intensity factor under plane strain conditions.

$$0.45 \leq \frac{a}{W} \leq 0.55 \tag{3}$$

$$a, (a\text{-}W), B \geq 2.4 \left(\frac{K_{IC}}{\sigma_y}\right)^2 \tag{4}$$

$$P_m \leq 1.10\, P_Q \tag{5}$$

where P_{max} is the maximum stress in the load-displacement curve.

Strain energy release rate G, is a measurement of the energy required to extend a crack per unit area. For a deformed specimen in a total elastic manner, it can be defined as:

$$G = \frac{1}{t}\frac{dU}{da}\Big|_{\Delta} \tag{6}$$

where t is the specimen thickness, dU is the energy input, da is the crack extension. In LEFM, catastrophic failure occurs when G reaches a critical value, G_c which can be calculated from the total fracture energy U:

$$G_c = \frac{U}{tW\emptyset} \tag{7}$$

where W is the specimen width and ∅ is a calibration factor depending on the testing configuration. As K_c under plane strain fracture, G_{Ic} is a true material constant, and some size criteria are required to fulfill. The subscribed I mean the fracture is under mode I, i.e., crack tip opening mode. There are other two fracture modes, i.e., mode II in plane-shear and mode III out-of-plane shear. Under mode I, K_{Ic} and G_{Ic} has its relationship as shown below.

$$G_{IC} = \frac{K_{IC}^2}{E} \text{ For plane stress} \tag{8}$$

$$G_{IC} = \frac{\left(\frac{K_{IC}^2}{E}\right)}{(1-\nu^2)} \text{ For plane strain} \tag{9}$$

where E is the Yong's Modulus and ν the Poisson's ratio.

Both K_{Ic} and G_{Ic} have proven to consist of true material constants under plane strain fracture conditions. Some size requirements listed before should be satisfied to ensure that the critical specimen sizes are at least 50 times larger than the plane strain plastic zone size [33, 34].

Elastic–Plastic Fracture Mechanics (EPFM)

A limited amount of crack-tip ductility is compatible with LEFM, because as long as the global response remains linear, the crack-tip conditions are *K*-controlled. However, for solids that show significant global nonlinearity, generally associated with extensive crack-tip ductility when the size requirements for valid LEFM testing cannot be met, elastic–plastic fracture mechanics (EPFM) may be more appropriate. This is often the case for the tougher polymers, given that the thickness of molded specimens is limited by processing constraints. Nowadays, the most successful methods for EPFM are the J-integral method and the Crack tip opening displacement (CTOD) method [35, 36].

J-Integral Method

The J-integral method is based on the path-independent contour integral introduced by Rice. The basic formula for J-integral is:

$$J = \int_{\Gamma}(W dy - T\,\frac{\partial u}{\partial x}\,ds) \tag{10}$$

where W is the strain energy density function in the direction normal to the crack line, T the traction vector, u the displacement vector, and Γ the curve surrounding the crack tip, which can take any path from one crack surface to the other outside the fracture process zone. Therefore, the value of J is path-independent, which means that the same values can be obtained for all choices of Γ.

Landes & Begley were among the first to invoke the energy release rate definition of J-integral from experimental measurement through using multiple specimens with different crack lengths. In their way, a critical value of J could be evaluated by:

$$J = -\frac{1}{B}\,(\frac{\partial U}{\partial a})_x \tag{11}$$

where U is the total strain energy absorbed by the specimen, which can be calculated from the area under a given load-displacement, curves as shown in Figure 8. B is the thickness of the sample. It was observed that the crack initiation occurred at the same value of J for the samples with different crack lengths. Therefore, they suggested using a critical value of J as a fracture criterion [37, 38].

Later, according to Sumpter & Turner, the equation mentioned above can be extended and re-written as:

$$J = J_e + J_p \tag{12}$$

$$J = \frac{1}{B}\int_0^P \left(\frac{\partial \nabla}{\partial a}\right)_P dP = \frac{\eta_e U_e}{B\,(W-a)} + \frac{\eta_P U_P}{B\,(W-a)} \tag{13}$$

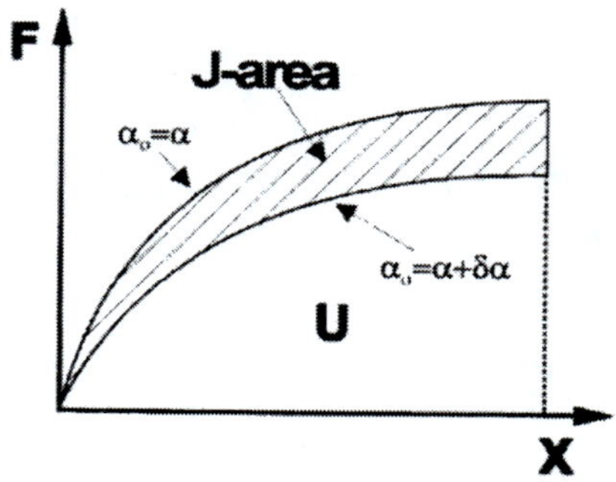

Figure 8. Schematic of J-integral determination.

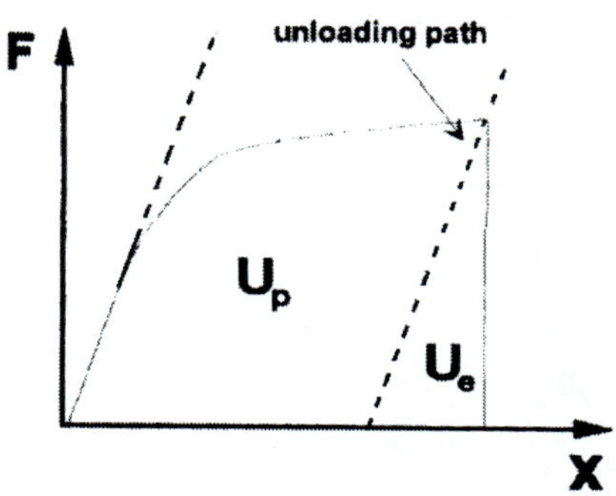

Figure 9. Energy used for J-integral determination.

J_e and J_P are the elastic and plastic terms of the J-integral related to the elastic and plastic components of the strain energy U_e and U_P, respectively, as shown in Figure 9 [35].

There are two techniques used for J-integral determination, i.e., single and multi specimens. In the multiple specimen method, a set of nominally identical specimens are loaded to different displacements to make various crack extensions, then unloaded and cooled to a low temperature and then broken up. The initial crack length and stable crack growth can be determined by optical measurements for each specimen. Then a series of data points of J- Δa can be plotted. In the single specimen method, the specimen is periodically unloaded by 10% of the input load. From the unloading compliances, the crack lengths and thus the crack extension Δa is calculated. This method is difficult to control and measured. Thus, the multi specimen technique is preferred for most works [39, 40]. To estimate the critical value J_{Ic} from the plotting of J against crack growth Δa, nowadays there are two standards, ASTM E 813-81 and ASTM E813-87, 89, as shown in Figure 10.

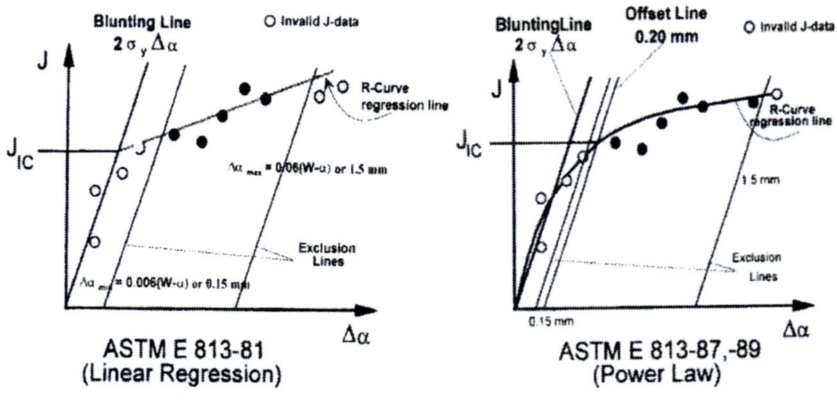

Figure 10. Schematics of data reduction according to ASTM 813-81 and -87; 89 standards, respectively.

Both of those standards were designed initially for metallic materials. Due to the unique nature of polymers and composites, the J-integral methods designed for metals may not be completely appropriate. Many studies have been done to transfer this method from ductile metals to polymeric materials and apply this toughness evaluation method to study the structure-toughness relationships for polymers and polymer-based composites. It shows that the J-integral approach is a useful and applicable tool to determine the fracture resistance of ductile and toughened polymers, which have low strength and high fracture toughness [41].

However, there are remaining some controversial aspects of the J-integral method when applying to polymeric materials. The materials with high resistance against tearing may not deliver satisfactory results because crack-growth due to physical tearing may not be distinguishable from the crack tip blunting process. It is recognized that it may not be proper to use a blunting line to determine the critical value of J-integral from the Jr-curve for some ductile polymers or polymer blends. For example, Hashemi & Williams found that the blunting line and the J_R-curve overlap for LLDPE tested at room temperature. In other instances, a pronounced nonlinear relationship between J and the amount of crack extension can also make it impossible to determine the critical value J_{IC}. To obtain a valid value of J_{IC}, a general size requirement should be fulfilled [42, 43].

$$W\text{-}a,\ B,\ a > 25\,\frac{J_{1C}}{\sigma_y} \tag{14}$$

Crack Tip Opening Displacement

The crack tip opening displacement criterion was proposed for the occurrence of crack blunting due to plastic deformation before crack initiation. The relationship between the crack tip opening displacement, δ_c, and energy release rate, G_C or J_C can be established by applying the Irwin model, Wells' concept, and Dugdale's solution. Burdekin and Stone used models as mentioned above to indicate that CTOD can be expressed as follows:

$$\delta_c = \frac{8a\sigma_y}{\pi E}\ln\left[\sec\frac{\pi\sigma}{2\sigma_y}\right] \cong \frac{\pi\sigma^2 a}{E\sigma_y} \cong \frac{K_C^2}{E\sigma_y} \quad (15)$$

CTOD can be employed to measure fracture toughness under both brittle and ductile fracture and for the ductile-brittle transition region. Usually, applications of CTOD are focused on the "yielding fracture mechanics" regime, i.e., between linear elastic fracture behavior and fully plastic behavior. It can describe very well in this transition region from elastic to plastic fracture mechanics, where the failure always occurs by plastic instability [44].

Post Yield Fracture Mechanics

Essential Work of Fracture Method

The essential work of fracture method (EWF) has been used to examine fracture characterization of ductile polymers, related blends, and composites. The EWF is the total work of fracture (W_f) which is calculated from the area of the load – displacement curves. According to Eq.16, W_f consists of two components of W_e and W_p that the essential work of fracture (W_e) related with the inner fracture process zone (FPZ), and the non-essential work of fracture (W_p) that corresponds with the plastic work done in the outer process zone (OPZ) [45].

$$W_f = W_e + W_p = w_e \times lt + w_p \times \beta t l^2 \; which \; \frac{W_f}{lt} = w_f = w_e + \beta w_p \times l \quad (16)$$

W_e, W_e, and W_p are total specific work of fracture, specific fracture work, and specific plastic work, respectively, β is the shape factor related to the form of the outer plastic zone and t is the specimen thickness. According to the energy-partitioned method recommended by Karger-Kocsis, the total work of fracture is separated into two components: the work for yielding the ligament region (w_y), and the energy required for necking and tearing ligament area (w_n).

$$W_f = W_y + W_n \tag{17}$$

According to Eq. (16) and (17), the EWF formulation can be rewritten as follows:

$$w_f = \left(w_{e,y} + \beta_y w_{p,y} l\right) + \left(w_{e,n} + \beta_n w_{p,n} l\right) \tag{18}$$

$w_{e,y}$ is the specific essential yielding-related work of fracture, $w_{e,n}$ is the specific essential necking, and tearing work, $w_{p,y}$ is the volumetric energy dissipated during yielding and $w_{p,n}$ is the dissipated plastic work during necking and tearing. The double-edge notched tension (DDENT) specimens to evaluate the essential work of fracture is shown in Figure 11 [2, 46].

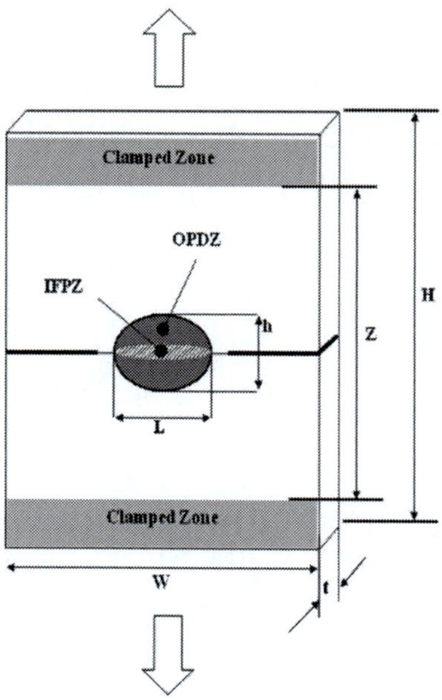

Figure 11. Double-edge-notched tensile (DENT) specimen used for EWF tests, showing the different energy dissipation zones involved.

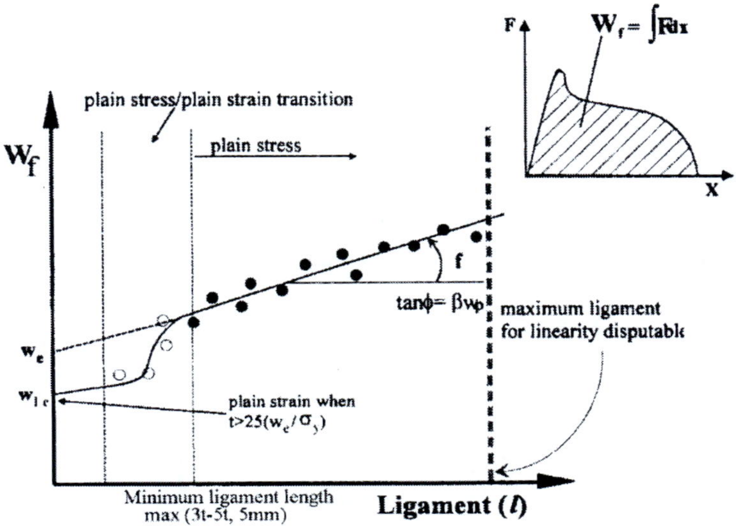

Figure 12. Determination of EWF parameters according to ESIS standards.

In the experimental section, a serial of specimens with different ligament lengths should be notched and tested to fracture and thus a corresponding serial of the total fracture energy can be obtained, which can be calculated from the load-displacement curves. According to the above equation, if plotting w, against l, there should exist a linear regression. The specific essential work of fracture and the specific non-essential work of fracture can be obtained from the intercept at Y-axis with zero ligament length and the slope of the line, respectively, as illustrated in Figure 12. Here it is supposed that β is independent of the ligament length, which should satisfy some basic requirements, i.e., the ligament should be fully yielded before the crack propagation starts and all the load-displacement curves have a geometric similarity [4, 47].

Polypropelene/Clay Nanocomposite

Organoclay or organically modified nanoclay/layered silicates is one of the most widely used reinforcing nanofillers to fabricate polymer nanocomposites. In the polymer matrix, polypropylene is considered one of the most commonly used thermoplastic polymers. Therefore, fabrication and investigation of polypropylene/clay nanocomposite properties are very interesting topics of research. This approach has been well developed for polypropylene-based systems; the focus of most industrial research and development has been on the formation of nanocomposites by melt-mixing or compounding processes because this is generally more attractive than in situ polymerization due to better commercial feasibility and lower cost [48]. Furthermore, compared to other methods, the melt blending method is more simple, economical, and thus more feasible for mass production of polymer/clay nanocomposites. It is also the most compatible method with the current industrial machines, such as extrusion and compression/injection molding. The clay is naturally a hydrophilic material, which has different polarity to PP. Therefore, a compatibilizer such as polypropylene-graft-maleic anhydride (PP-g-MA) is needed to facilitate the dispersion of clay layers in PP. The interaction between PP-g-MA and clay layers helps to expand clay galleries; as a result, the interaction between clay layers weakens. In many approaches, first, a masterbatch with high clay loadings of PP-g-MA is prepared, and then mixed with PP to form a composite with the required clay content [49, 50]. In this process, when the PP-g-MA intercalated clay contacts with PP under processing, due to strong shear field, delamination of the clay layers may be facilitated if there is a signicant enough interaction strength existing between the PP and PP-g-MA to form a molecularly dispersed clay nanocomposite (Figure 13).

Nam et al. [51] prepared PP nanocomposites using PP-g-MA (0.2 wt% MA) and nanoclay in a melt extrusion method using a twin-screw extruder at 200°C. The prepared nanocomposites with 2 wt% clay showed an exfoliated structure; with 4 wt% clay, a disordered intercalated structure; and with 7.5 wt% clay content, the nanocomposite showed an ordered intercalated structure.

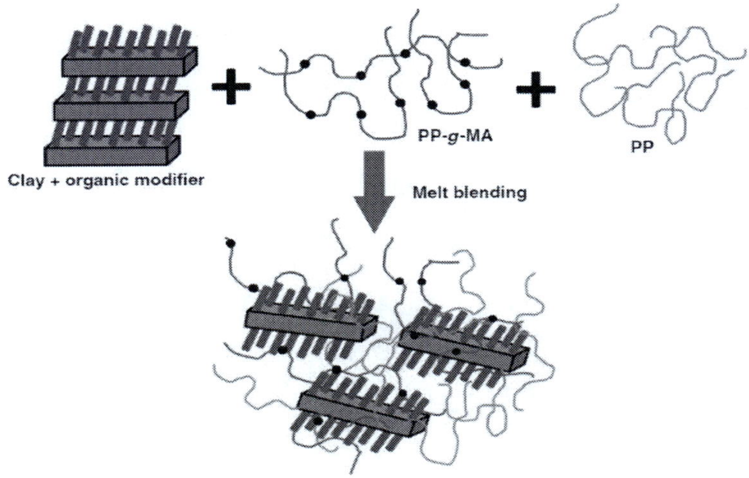

Figure 13. Schematic representation of the melt compounding method used to prepare PP/Clay nanocomposite in the presence of PP-g-MA compatibilizer.

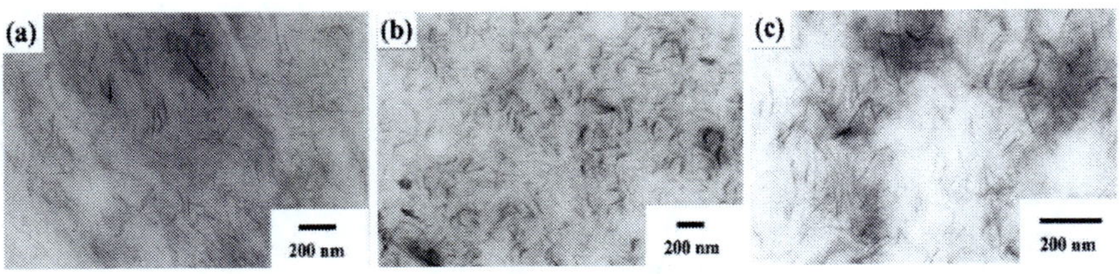

Figure 14. TEM micrograph of (a) PPC2 (b) PPC4 (c) PPC7.5 wt % blends.

Zheng et al. [52] studied the effect of clay on the polymorphism of PP in PP/clay nanocomposites. They found that, in both extruded PP and PP nanocomposite samples, only α-form crystallites were formed. They also prepared the samples of PP and PP/Clay nanocomposite under compression. It was found that, in the pure PP samples that were compressed at 200°C and crystallized at various temperatures and times, the maximum amount of β crystallites were formed. This indicates that the crystallization temperature has a significant effect on the formation of β-phase crystallites in neat PP. In the compressed films of PP/Clay nanocomposite, prepared under the same thermal and processing conditions, no β-phase crystals were detected. This reveals that the clay galleries significantly inhibit the formation of β-phase PP crystallites in the nanocomposites.

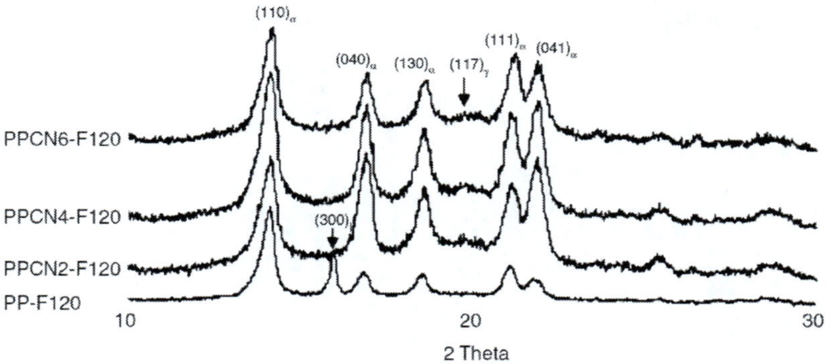

Figure 15. XRD patterns of the neat PP and PPCN film samples prepared by isothermal crystallization at 120°C on a hot press. Here, PPCN2-F120 identified the PP sample with 2 wt% clay and is annealed at 120°C on a hot press.

The thermal behavior of a polymer material and polymer nanocomposite are studied by thermogravimetric analysis (TGA) and differential scanning calorimetry (DSC). Mohan et al. [53] showed that the increase in clay content leads to a decrease in crystallization temperature from 169 to 163°C. The melting temperature of PP/clay nanocomposite prepared with 15A clay particles modified by an intercalating ammonium ion base is greater than the melting temperature of PP/clay nanocomposite prepared with Na montmorillonite particles modified by an intercalator amine base.

Thermogravimetric analyses in air showed a shift of the weight loss curves of nanocomposites towards higher temperatures with respect to neat polymer [Modesti]. In other words, the nanoclays brought an increase in thermal stability in the oxidizing atmosphere of PP. In an inert atmosphere, we registered a marked shift of the onset temperature, which increases, in the better case, by about 200°C; the effect of different processing conditions is not so evident (Table 1). In this Table, the number after the letters in the sample name refers to the weight % of filler: for example, PP-LH-5 refers to a PP-based sample processed with a low-temperature profile, high screw speed, filled with 5 wt% of nanoclay.

Table 1. Thermogravimetrical data for LH samples

Sample	Air		Nitrogen	
	$T_{20\%}$	$T_{50\%}$	$T_{20\%}$	$T_{50\%}$
PP-LH	341	377	411	435
PP-LH-3.5	368	406	416	436
PP-LH-5	371	407	436	443
PPgMA-LH	321	359	436	454
PPgMA-LH-3.5	382	424	446	450
PPgMA-LH-5	383	425	448	452

$T_{20\%}$ and $T_{50\%}$ are the temperatures in°C at which occurs the 20% and 50% weight losses, respectively.

Maiti et al. [54] measured the dynamic behavior of PP nanocomposites. They found that the storage modulus G′ of a nanocomposite with 4 wt% of clay is higher than that for pure PP samples. In addition, with an increase of T_c, G′ increases, where the enhancement is maximal (30 %) for a PP nanocomposite with 4 wt% clay content. For further increases in the clay content, the G′ decreases and is 13% (Table 2). They attributed that the greater increase of the storage modulus is related to the extent of clay dispersion.

Table 2. Storage Modulus of PP-MA and PPCNs at Different Temperatures

system	T_c/°C	$G' \times 10^{-8}$/Pa	% increase
PP-MA	70	2.92	9.9
	130	3.21	
PPCN2	70	4.79	
	130	4.50	
PPCN4	70	5.16	30.6
	130	6.74	
PPCN7.5	70	7.49	13.3
	130	8.49	

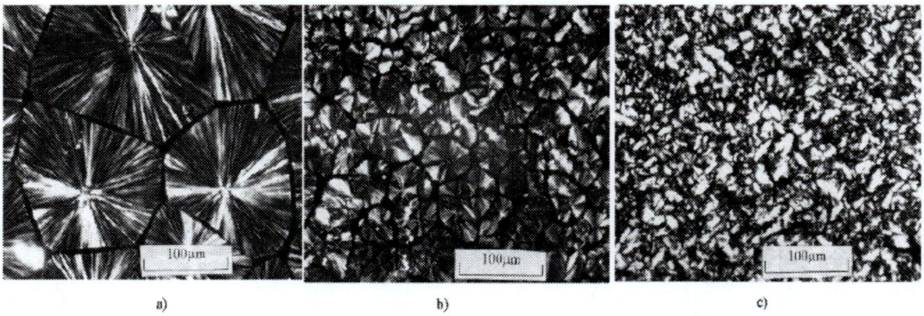

Figure 16. Micrographs of (a) Pure PP, (b) PP/4.6 wt%, and (c) PP/8.4 wt% clay.

Table 3. Mechanical Properties of the PP/Clay Nanocomposites

PP/Clay Composites	Tensile Strength (MPa)	Tensile Modulus (MPa)	Elongation (%)	Flexural Strength (MPa)	Flexural Modulus (MPa)
PP	32	1450	28.2	32.3	1390
PP/clay 6A	35.8	2028	7.7	39.9	1785

The clay alone leads to a greater rate of nucleation and smaller spherulites, for intercalated nanocomposites. Figure 16 clearly shows the decrease in spherulite size by increasing the rate of clay.

The enhancement of the mechanical properties of polymer-clay nanocomposites is mainly dependent on the dispersion state of clay, the reinforcing effect, and the interfacial interaction between the clay and the polymer matrix. Kodgire et al. [55] investigated the effects of nanoclay on the mechanical properties of the PP/clay composites. According to Table 3, the results show that the PP/clay nanocomposites exhibited improved mechanical properties at a clay loading of 4% by weight. The tensile modulus increased by about 40%, whereas the tensile strength increased by 10%. The flexural modulus also increased by about 30%, whereas the flexural strength increased by 25%.

The fracture behavior of polymer nanocomposites (PNCs) based on polypropylene with organo-modified clays (2 wt %) and different coupling agents was studied by Bureau et al. Three commercially available MAgPPs with a low amount of grafted MA were used as coupling agents: one with a low molecular weight (Mw = 56 kg/mol), defined as LMW, one with an intermediate molecular weight (Mw = 195 kg/mol), defined as IMW and one with a high molecular weight (Mw = 330 kg/mol), defined as HMW. The designation of the materials studied is presented in Table 4.

Table 4. Designation of the materials

Material	Coupling agent	Clay
PP	–	–
PP/clay	–	15A
L-PNC1	LMW	15A
L-PNC2	LMW	20A
I-PNC1	IMW	15A
I-PNC2	IMW	20A
H-PNC1	HMW	15A
H-PNC2	HMW	20A

They concluded that PP and PP/clay specimens exhibit a brittle behavior, where most, and in the case of PP/clay, even all, of the strain energy is absorbed during crack initiation. In the case of L-PNCs, I-PNCs, and H-PNCs, well-developed crack initiation and crack propagation regions are observed on the load–displacement curves. In addition, fractographic observations were made to understand the influence of PNC nano / microstructure on the fracture behavior, especially the mechanisms that lead to their fracture. For PP/clay, fracture surfaces feature voids separated by short ligaments (Figure 17a). These voids were formed at large clay aggregates (Figure 17b) and grew under the imposed deformation until they coalesced with neighboring voids. This coalescence left ligaments of the unbroken matrix that bridged the crack process zone and eventually failed. Higher magnification observation of PP/clay fracture surface shows that clay aggregates were large, above 1 μm in size, and very low particle-matrix adhesion prevailed (Figure 17c). The presence of such large defects (stress concentrators) poorly coupled to a low toughness matrix explains the brittle behavior obtained for this material.

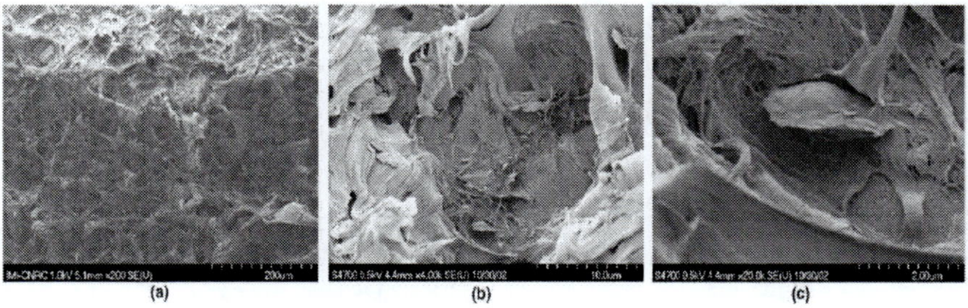

Figure 17. Fractographic observations of PP/clay at (a) low (200), (b) intermediate (4000), and (c) high (20,000) magnification.

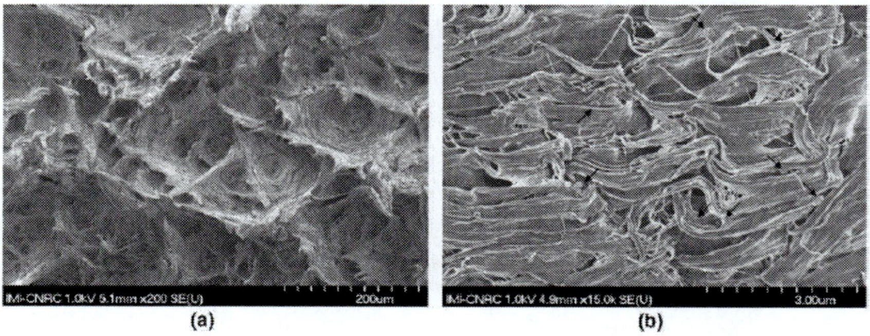

Figure 18. Fractographic observations of L-PNC1 at (a) low (200) and (b) high (15,000) magnification. Arrows indicate clay particles.

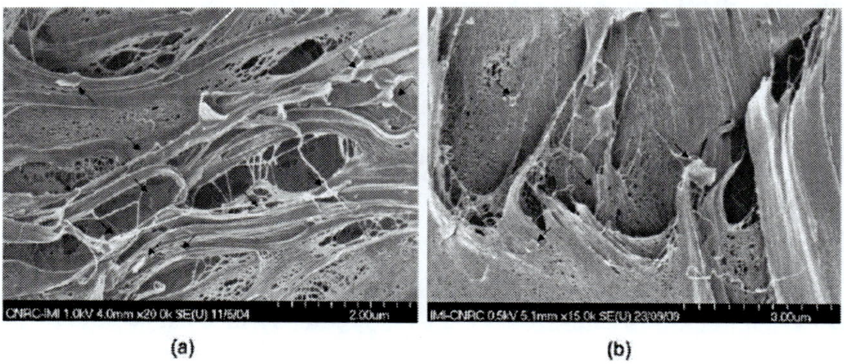

Figure 19. Fractographic observations of (a) I-PNC1 at high (20,000) magnification (b) H-PNC1 at high (25,000) magnification. Arrows indicate clay particles.

L-PNCs specimens (Figure 18) failed essentially in a similar fashion, but fracture surfaces feature fewer, larger voids with considerably elongated ligaments. Clay particles were also smaller and appeared attached to the deformed matrix, in either ligaments or voids. Similar fewer, larger voids with considerably elongated ligaments were observed on the fracture surfaces of I-PNCs and H-PNCs. The matrix was deformed in small-elongated fibrils, between which important microvoid formation was observed. Such microvoids are shown for L-PNCs, I-PNCs, and H-PNCs (Figs. 18b, 19 (a), and (b)). At least in the case of I-PNCs (Figure 19 (a)), for which it was very clear, very small high aspect particles remained attached to the elongated fibrils observed. Similar observations could sometimes be made for H-PNC (Figure 19 (b)), although to a lesser extent.

Rajesh et al. [56] studied the effect of processing parameters (Injection flow rate (Q), Holding pres-sure (HP), Backpressure (BP), Screw speed (SS)) on mechanical properties of PP /clay nanocomposites fabricated by injection molding. . Factors (injection molding parameters) and levels selected in this study shows in Table 5. The various samples obtained that way is the following: PP-H and NC-H, respectively, for the PP and PP nanocomposites which were injection-molded with all set-up parameters at high levels; PP-L and NC-L, respectively, for the PP and PP nanocomposites which were injection molded with all set-up parameters at low levels.

Table 5. Factors (injection molding parameters) and levels selected in the DOE

	Factors				
	Assigned test level	Injection flow rate (Q) [cm^3/s]	Holding pres-sure (HP) [bar]	Back pressure (BP) [bar]	Screw speed (SS) [rpm]
Assigned set-up level	High level (H)	50	350	65	90
	Low level (L)	25	250	35	50

They confirmed that the clay dispersion is influenced by the injection molding conditions (Figure 20). This difference of dispersion degree is sufficient to lead to variations in mechanical properties. The values of tensile properties (Young modulus, yield stress, stress and elongation at break), and Charpy impact strength are reported in Table 6.

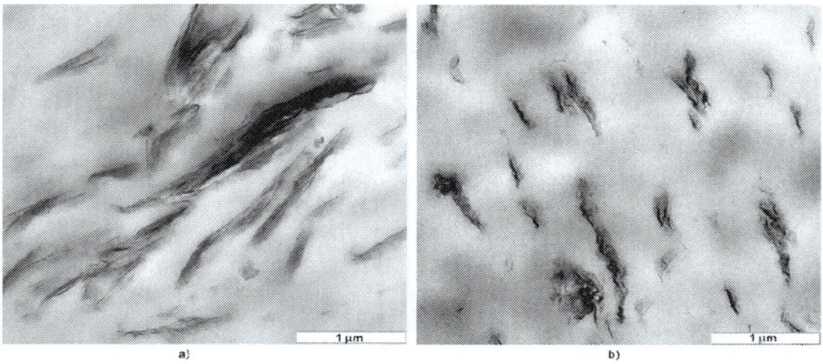

Figure 20. TEM images surface of (a) NC-L and (b) NC-H nanocomposite.

Table 6. Tensile and Charpy impact properties of PP and PP nanocomposites

Material	Young modulus [MPa]	Yield stress [MPa]	Tensile el-ongation at break [%]	Stress at break [MPa]	Impact strength – unnotched [kJ/m^2]	Impact strength – notched [kJ/m^2]
PP-L	1280 ± 20	28.9 ± 0.3	616 ± 15	35.2 ± 0.6	117 ± 8	2.4 ± 0.2
PP-H	1254 ± 23	28.8 ± 0.3	613 ± 12	34.7 ± 0.5	120 ± 9	2.3 ± 0.1
NC-L	1617 ± 54	35.3 ± 0.6	100 ± 22	8.7 ± 6.1	52 ± 4	4.3 ± 0.9
NC-H	1687 ± 45	35.6 ± 0.6	103 ± 23	12.9 ± 5.2	64 ± 8	3.9 ± 0.9

Compared to neat PP (PP-L and PP-H), the increase of the Young modulus for NC-L is about 25%, whereas the increase for NC-H is slightly higher (30% compared to PP). The dispersion degree does not significantly modify the yield stress. The influence of the clay dispersion degree is much more visible on the stress at break. The higher dispersion degree in NC-H nanocomposite tends to increase significantly (+50%) the stress at break. Whereas the elongation at break is usually expected to be influenced by the dispersion state, the difference between NC-L and NC-H is here negligible, considering the decrease of ductility when the clay platelets are added into the PP matrix and the standard deviations (experimental data dispersion). A similar trend is observed for the unnotched impact strength (e.g., toughness decrease upon adding of nanoclay into PP). Conversely, in the case of notched impact strength, the nanocomposites NC-L and NC-H have a higher toughness than PP. This is because notched impact behavior is controlled to a greater extent by factors affecting the propagation of fracture due to stress concentration at the notch tip. The presence of dispersed clay tactoids may restrict the propagation leading to higher toughness, as already observed in PP/carbon nanotube nanocomposites. Concerning the impact strength, the toughness of the better-dispersed NC-H nanocomposites tends to be higher (+20%) when measured on unnotched samples.

Polypropelene/Carbon Nanotube Nanocomposite

Carbon Nanotubes (CNTs) are cylindrical shells made, in concept, by rolling graphene sheets into a seamless cylinder. CNTs exist as either single-wall nanotubes (SWNTs) or multi-walled nanotubes (MWNTs). The SWNT consists of a single graphene sheet, a planar array of benzene molecules, involving only hexagonal rings with double and single carbon-carbon bonding. The choice of the rolling axis relative to the hexagonal network of the graphene sheet and the radius of the closing cylinder allows for different types of SWNTs, which vary from insulating to conducting. According to the rolling angle of the graphene sheet, CNTs have three chiralities: armchair, zigzag and chiral one. The tube chirality is defined by the chiral vector, $C_h = na_1 + ma_2$ (Figure 21), where the integers (n, m) are the number of steps along the unit vectors (a_1 and a_2) of the hexagonal lattice [57]. Using this (n, m) naming scheme, the three types of the orientation of the carbon atoms around the nanotube circumference are specified. If $n = m$, the nanotubes are called "armchairs." If $m = 0$, the nanotubes are called "zigzag." Otherwise, they are called "chiral." The chirality of nanotubes has a significant impact on their transport properties, particularly their electronic properties. For a given (n, m) nanotube, if $(2n + m)$ is a multiple of 3, the nanotube is metallic. Otherwise, the nanotube is a semiconductor. Each MWCNT contains a multi-layer of graphene, and each layer can have different chiralities, so the prediction of its physical properties is more complicated than that of SWCNT [58, 59].

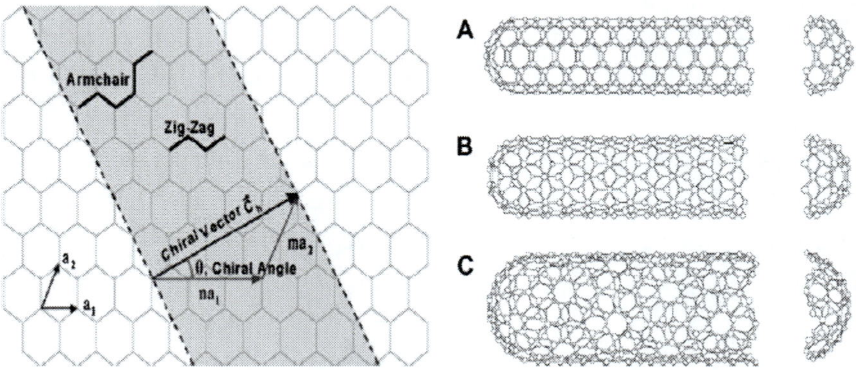

Figure 21. Schematic diagram showing how a hexagonal sheet of graphene is rolled to form a CNT with different chiralities (A: armchair; B: zigzag; C: chiral).

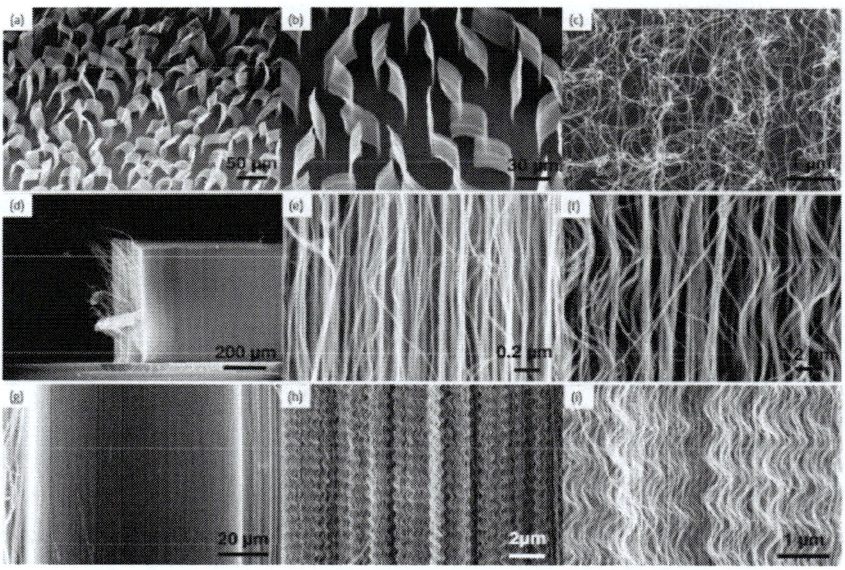

Figure 22. (a) and (b) SEM images of CNT thin sheets array. The ~ 100μm height sheets were grown from 0.2 μm wide and 40 μm long catalyst thin films. (c) CNTs in the thin sheet. (d) and (e) super-aligned MWNT array. (f) Straight and waved MWNTs in an array. (g) to (i) SEM images of a MWNT array with wavy structure at different magnifications.

A variety of techniques have been developed to produce CNTs. MWNTs were synthesized and observed a few decades ago. In 1993, Iijima et al. and Bethune et al. independently reported the synthesis of SWNTs. Primary synthesis methods for single-wall and multi-walled CNTs include arc discharge, laser ablation, gas-phase catalytic growth from carbon monoxide, and chemical vapor deposition (CVD) from hydrocarbons. Synthesis methods such as arc discharge, laser ablation, and certain types of CVD with floating catalysts produce nonaligned, entangled ropes of nanotubes. The diameter of a CNT is nanometers, while its length can be up to several millimeters. Long CNTs usually can remain straight when they are in oriented structures [57, 60].

CNTs have various tubule morphologies and appear in different shapes such as straight, waved, coiled, and branched. Each shape has its special properties and potential applications. CNTs in different shapes will have significant effects only when they can be produced with uniform properties in large quantities.

Although there is no consensus on the exact mechanical properties of CNTs, theoretical and experimental results have shown unusual mechanical properties of CNTs with Young's modulus as high as 1.2 TPa and tensile strength of 50–200 GPa. These make CNTs the strongest and stiffest materials on earth. In addition to the exceptional mechanical properties associated with CNTs, they also possess other useful physical properties and Table 7 summarizes these properties [57, 61].

Table 7. Physical properties of different carbon materials

	C material				
Property	Graphite	Diamond	Fullerene	SWCNT	MWCNT
Specific gravity (g/cm^3)	1.9–2.3	3.5	1.7	0.8	1.8
Electrical conductivity (S/cm)	4000^p, 3.3^c	10^{-2}–10^{-15}	10^{-5}	10^2–10^6	10^3–10^5
Electron mobility ($cm^2/(V\,s)$)	2.0×10^4	1800	0.5–6	$\sim 10^5$	10^4–10^5
Thermal conductivity (W/(m K))	298^p, 2.2^c	900–2320	0.4	6000	2000
Coefficient of thermal expansion (K^{-1})	-1×10^{-6p} 2.9×10^{-5c}	$(1\sim3)\times10^{-6}$	6.2×10^{-5}	Negligible	Negligible
Thermal stability in air (°C)	450–650	<600	~600	>600	>600

p: in-plane; c: *c*-axis.

Many research efforts have been made to understand their structure-property relationship and find useful applications in different fields. These efforts have become more pronounced after the realization of CNT fabrication on an industrial scale with lower costs at the beginning of the 21st century. According to the specific application, CNT/polymer nanocomposites can be classified as structural or functional composites. For the structural composites, the unique mechanical properties of CNTs, such as the high modulus, tensile strength, and strain to fracture, are explored to obtain structural materials with much improved mechanical properties. As for CNT/polymer functional composites, many other unique properties of CNTs, such as electrical, thermal, optical, and damping properties along with their excellent mechanical properties, are utilized to develop multi-functional composites for applications in the fields of heat resistance, chemical sensing, electrical and thermal management, photoemission, electromagnetic absorbing and energy storage performances, etc [62].

The combination of these exceptional mechanical properties, along with the low density, high aspect ratio, and high surface area, make CNTs an ideal candidate for reinforcement in composite materials. Both SWCNTs and MWCNTs have been utilized for reinforcing thermosetting polymers, such as epoxy, polyurethane, and phenol-formaldehyde resins, as well as thermoplastic polymers, including polyethylene, polypropylene, polystyrene, nylon, and so on. The CNT reinforced nanocomposites can be considered particulate composites with filler dimensions on the nanometer scale and a high aspect ratio. Unlike the microscopic particulate composites, the mechanical properties of nanocomposites depend strongly on the dispersion state of nanofillers, apart from the properties of fillers and matrix material. In addition to dispersion, other major requirements need to be satisfied for effective reinforcement of CNTs in composites [11]: they include a high aspect ratio, alignment, and interfacial interactions between CNT and polymer matrix [58].

Bao et al. [63] determined the tensile and impact behaviors of the PP/MWNT nanocomposites prepared by dynamic packing injection molding. Tensile measurements were performed at room temperature under a crosshead speed of 50 mm/min. Both tensile strength and impact toughness were found to increase simultaneously when the MWNTcontent reached 0.6 wt%. These improvements in mechanical properties were attributed to uniform dispersion and the possible orientation of nanotubes induced by shear stress.

Yetgin [64] investigated the effect of MWCNT amount on mechanical properties of PP/clay nanocomposites with 5 wt% nanoclay. They found that the mechanical test results showed that the incorporation of MWCNTs increased the tensile strength, flexural strength, and modulus of elasticity. At the same time, it decreased the impact strength and elongation at break values of PP/MWCNT nanocomposites. Table 8 represents the results of the mechanical properties of PP and PP/MWCNT nanocomposites.

In addition, it was obtained that the melting behavior and melting temperature(Tm) of PP did not change much with the addition of MWCNT. The melting temperature range of PP and its nanocomposites was in the range of 150–152◦C. However, the degree of crystallinity (X_c%) values of nanocomposites varied depending on the MWCNT amount. As seen in Table 9, the X_c value of PP increased about 5% with the MWCNT amount increased up to 0.3 wt%, then it slightly reduced with the further addition of MWCNT.

The flexural and morphological properties of PP/MWCNT composites analyzed by Patti et al. In detail, formulations containing 0.5%, 1%, and 1.5% non-functionalized and amino-functionalized MWCNTs were considered. The results show that the flexural modulus increases approximately up to 17.4% and 36.3% in the presence of 0.5% by volume of non-functionalized (NF) and amino-functionalized (NH2) carbon nanotubes, respectively. However, further increases of the filler content always induce slight reductions of the same parameter, probably due to the clusters or agglomerates. Similarly, it is noted an increase, with respect to the neat matrix, of the averaged flexural strength approximately equal to 15.1% for compounds containing 0.5 vol% of nanotubes, but in this case, the effect is not influenced by the surface functionalization.

Table 8. Results of mechanical properties of PP and PP/MWCNT nanocomposites

Samples	Tensile strength (MPa)	Tensile modulus (MPa)	Elongation at break (%)	Impact strength (kJ/m²)	Flexural strength (MPa)	Flexural modulus (MPa)
PP	29.76	950	150	9.23	26.4	771
PP/0.1MWCNT	30.27	1020	100	9.05	27.8	815
PP/0.2MWCNT	32.64	1168	82	8.71	30.2	886
PP/0.3MWCNT	33.35	1214	39	8.24	32.2	924
PP/1MWCNT	33.74	1289	28	7.85	33.6	1023
PP/2MWCNT	35.26	1378	19	7.41	35.7	1148

Table 9. Crystallization and melting parameters of PP and PP/MWCNT nanocomposites determined by DSC analysis

Samples	Cooling			Second heating		
	$T_{c,o}$ (°C)	$T_{c,p}$ (°C)	ΔH_c (J/g)	T_m (°C)	ΔH_m (J/g)	X_c (%)
PP	112.5	107.6	62.5	152.13	66.1	31.6
PP/0.1MWCNT	112.7	108.2	71.4	150.81	74.8	35.8
PP/0.2MWCNT	113.4	108.4	68.6	151.11	75.9	36.4
PP/0.3MWCNT	113.9	108.6	69.1	151.46	75.6	36.3
PP/1MWCNT	114.6	109.5	65.8	151.08	69.4	33.5
PP/2MWCNT	115.5	109.6	61.6	152.25	68.5	33.4

$T_{c,o}$, crystallization peak onset temperature; $T_{c,p}$, crystallization peak maximum temperature; ΔH_c, crystallization enthalpy; Tm, melting temperature; ΔHm, melting enthalpy (heat of fusion); X_c, degree of crystallinity.

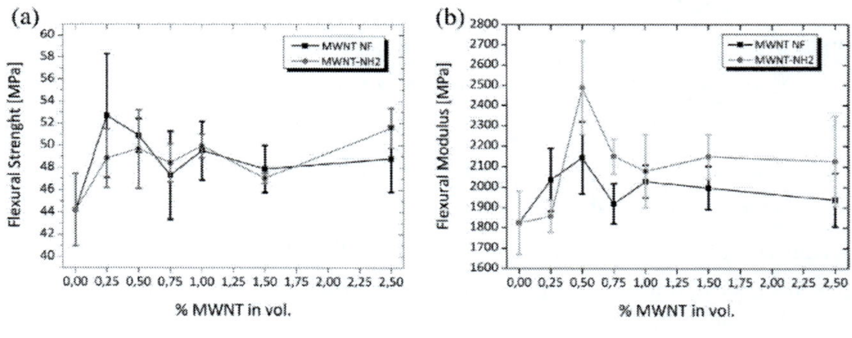

Figure 23. (a) Flexural modulus and (b) flexural strength as a function of the MWNT content.

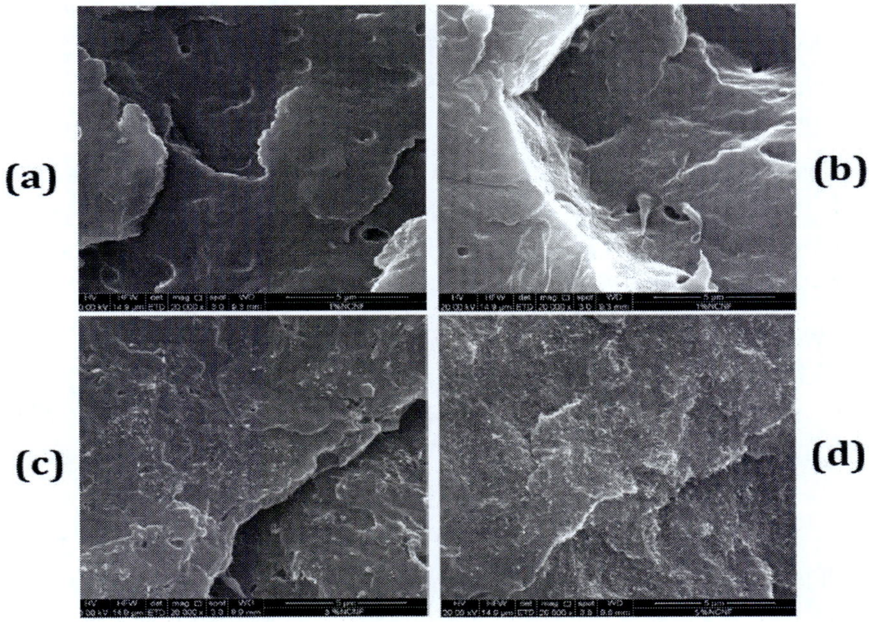

Figure 23. SEM micrographs of PP compounds containing (a) 0% (b) 0.5 vol%, (c) 1.5 vol% and (d) 2.5 vol% of MWNT-NF.

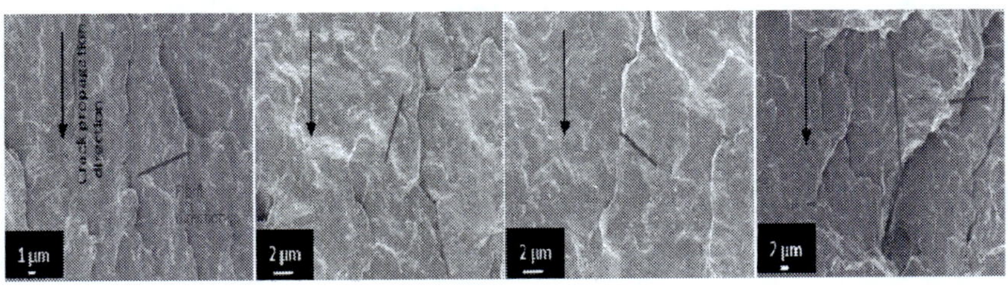

Figure 24. SEM micrographs of PP/ MWNT (1 wt %) nanocomposite impact tested at room temperature showing fibrils perpendicular/along the crack propagation direction.

According to Figure 24, with the exception of the neat PP micrograph showing typical features of a brittle fractured surface, in both cases of compounds with MWNT from the selected pictures, it is evident that nanotubes are mainly dispersed as bundles with size increasing with the carbon nanotubes content. In particular, at low filler content, MWNT appeared individually in the matrix or in the form of small aggregates.

Instead, at relatively high filler content, MWNT agglomerates are observed with larger dimensions.

The fracture mechanism of PP/f-MWNT nanocomposite was studied by SEM fractography and interfacial shear strength determination using Raman spectroscopy by Wang [65]. They concluded that the fracture surface of neat PP exhibits a cleavage type brittle appearance with unidirectional crack propagation direction. In contrast, the fracture surface of PP/f-MWNT nanocomposite becomes rougher and consists of several microcracks, which are possibly nucleated by CNTs, to lower the constraint in the matrix. The cracks propagated in a more tortuous path than in neat PP due to the reduced spherulite size. Fibril formation that consumed the fracture energy can be observed both in the plane and in out-of-plane. These fibrils were drawn not only along but also perpendicular to the crack propagation direction suggesting a more complex in-plane stress profile around the crack tip (Figure 24). Furthermore, shear-to-ruptured fibrils accompanied by the extension of the PP matrix can be found at the slow crack propagation zone and the stick-slip region. Such intense plastic deformation is attributed to the good matrix-CNT adhesion that allows effective load transfer at the CNT bridging cracks.

Polypropelene/Metal Oxide Nanocomposite

Titanium Dioxide

Titanium dioxide (TiO_2) exists in three different crystalline forms. They are rutile, anatase, and brookite. The most common form is rutile that is also the equilibrium phase at all temperatures. The metastable anatase and brookite phases get converted convert into rutile upon heating. Rutile, anatase, and brookite all contain six-coordinated titanium. Titanium is in octahedral coordination with O atom as TiO_6 units in all three forms. The different lattice structures arise due to the difference in arrangements of TiO_6 octahedra that is due to the difference in Ti-O bond length and angle. The unit cell of rutile and anatase is given in Figure 25 [66, 67].

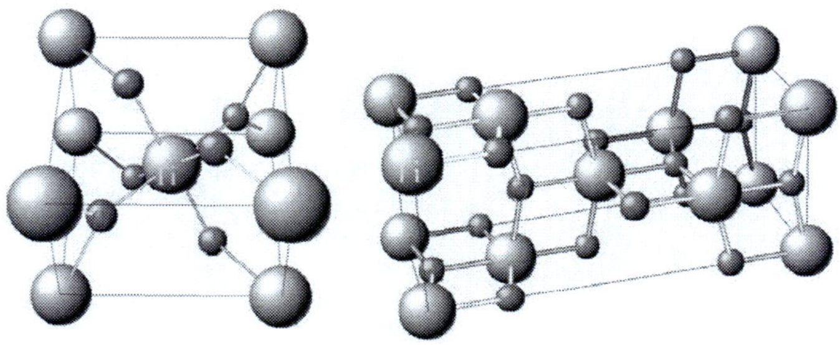

Figure 25. The unit cell of rutile and anatase crystal structure of TiO_2.

TiO_2 is used as a pigment to improve the white color of certain food materials such as dairy products and candy. It also lends brightness to toothpaste and some medications. It is suitable for use as a protective coating for many products like automobile parts and optical mirrors. Due to its refractive ability, it is a component of paints used to coat cars, boats, and airplanes. TiO_2 is also used in several of construction and building materials. In the plastic industry, TiO_2 is used as a coating to absorb UV light and make the material durable. TiO_2 is the most widely used white pigment due to its brightness and high refractive index ($n = 2.7$), in which it is surpassed only by a few other materials. Approximately 4 million tons of pigmentary TiO_2 are consumed annually in the world. When deposited as a thin film, it is an excellent reflective optical coating for dielectric mirrors [68].

Esthappan et al. [68] investigate the morphology and mechanical properties of the PP/TiO_2 fibers. They concluded that the tenacity of the PP fiber (Figure 26) is increased by the addition of titanium dioxide nanoparticles, indicating the reinforcing effect of TiO_2. In the case of filler reinforced fiber, the tenacity depends on the distribution and orientation of the filler. At 3 wt% of TiO_2, tenacity is increased by 72.69% indicates the good orientation of TiO_2 in the PP fiber. At lower concentration, tenacity is decreased, maybe due to the lack of good orientation of TiO_2 nanoparticles in PP fiber. Variation of elongation at break of the PP fibers with filler loading is shown in Figure 27. An increase of 15.79% is observed at of 1.5 wt% nano titanium dioxide. This indicates a decrease in the brittle nature of the fiber.

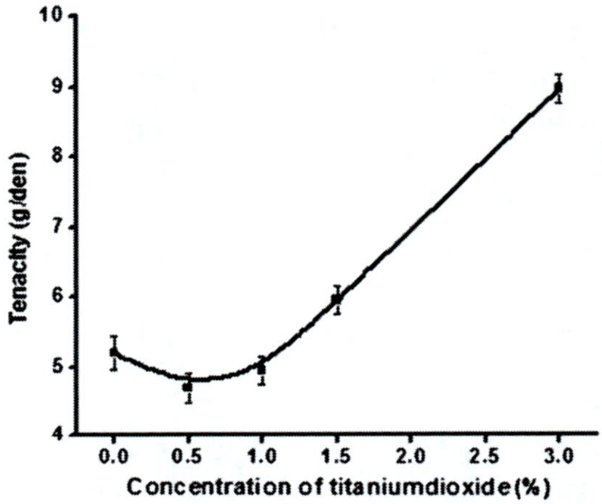

Figure 26. Effect of TiO_2 nanoparticles on the tenacity of PP fiber.

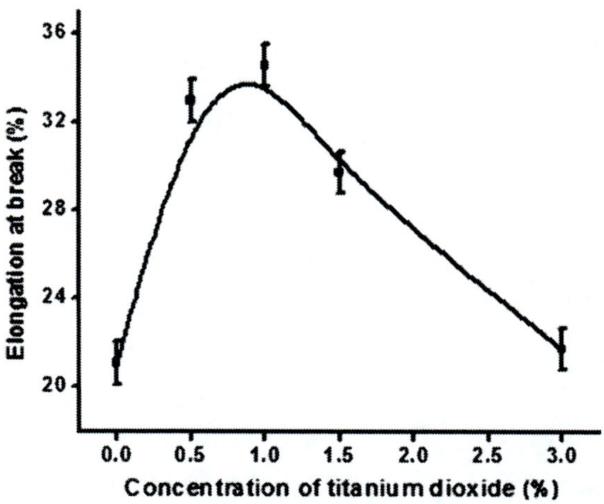

Figure 27. Effect of TiO_2 nanoparticles on the elongation at break of PP fiber.

Time to rupture of the fiber (Figure 28) increases reaches a maximum and then decreases at higher loadings. The maximum value is 46.2 s, at a loading of 0.5 wt%. Figure 6 shows the change in modulus with the concentration of nano-TiO_2. A substantial increase in modulus is observed at 3 wt% of TiO_2. The improvement in modulus of the fibers is related to the inherent stiffness and quality of the dispersion of TiO_2 and due to adhesion between the matrix and nanoparticles. The linear density of the fibers increased by 29.16% at 3 wt% of nano-TiO_2, as is evident from Figure 29. An increase in linear density indicates an increase in the fineness of the fiber.

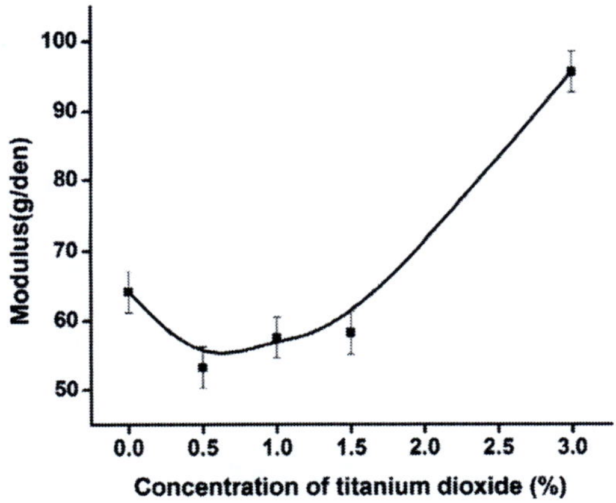

Figure 28. Effect of TiO_2 nanoparticles on the modulus of PP fiber.

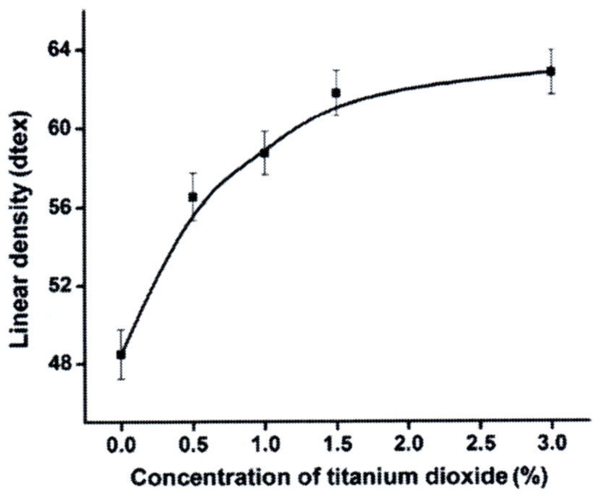

Figure 29. Effect of TiO_2 nanoparticles on the linear density of PP fiber.

This is also supported by SEM photographs. These results reveal that TiO_2 nanoparticles offer a substantial level of reinforcement to PP fibers.

The improved mechanical properties can be attributed to the effective matrix–filler interaction enabling load transfer from the polymer matrix to the TiO_2 nanoparticles.

Table 10. Differential scanning calorimetric analysis of PP/TiO2nanocomposite fibers

Sample name	T_c (peak crystallization temperature °C)	ΔH_c (J/g)	T_m (peak melting temperature °C)	ΔH_m (J/g)	Percentage crystallinity (%)
Neat PP	116.68	96.50	166.66	89.50	58.48
PP + 0.5% NTO	115.33	99.78	163.95	76.58	60.56
PP + 1.5% NTO	114.93	99.46	164.45	78.58	61.19
PP + 3% NTO	115.57	92.18	164.55	69.29	57.59

In addition, differential scanning calorimetry was used for studying the crystallization characteristics of polymers and their composites. The crystallization temperatures (T_c), the apparent melting temperature (T_m), and the corresponding enthalpies for the samples are shown in Table 10. Crystallinity increases with loading reaches a maximum at 1.5 wt% and then decreases. Neat fiber shows 58.48% crystallinity while 1.5 wt% TiO_2 filled fiber shows 61.197% indicating that TiO_2 nanoparticles are acting as nucleating agents for PP crystallization. These findings are most probably due to the nucleation role of nano-TiO_2 in the crystallization of PP, which results in a higher degree of crystallinity. But it can be inferred that larger aggregates prevent the crystal growth and hence reducing the crystallinity as the nano-TiO_2 content increases up to 3%.

Zinc Oxide

ZnO crystallizes in two main forms, hexagonal wurtzite, and cubic zinc blende. The wurtzite structure is most stable and common. The zinc-blended form can be stabilized by growing ZnO on substrates with a cubic lattice structure. In both forms, the zinc and oxide centers are tetrahedral. In addition to the wurtzite and zinc blende forms, ZnO can be crystallized in the rocksalt structure at relatively high pressures of about 10 GPa. Hexagonal and zinc blende forms have no inversion symmetry. Various lattice symmetry properties result in piezoelectricity of the hexagonal wurtzite and cubic zinc blende and pyroelectricity of hexagonal wurtzite. The wurtzite and zinc blende structure is given in Figure 30 [69, 70].

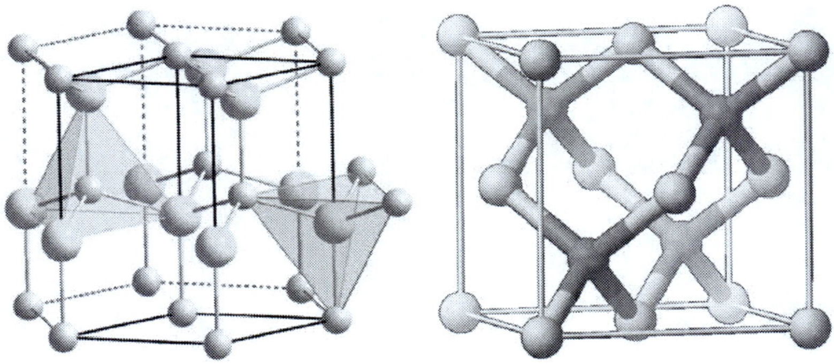

Figure 30. Unit cell structure of wurtzite and zinc blende of ZnO.

Nano-ZnO as a functional inorganic filler has been widely used in functional devices, catalysts, pigments, optical materials, cosmetics, and UV absorbers. Introduction of nano ZnO into polymers can enhance the mechanical properties of the polymers because of a strong interfacial interaction between polymers and nanoparticles, endow polymers with some other functional capabilities, such as photostabilization, protect matrix resins from environmental degradation, and increase the toughness. Moreover, its smaller diameter and relatively lower quantity of addition would have little influence on the transparency of the products [70, 71].

Altan et al. [72] investigated the tensile, morphological, and thermal properties of the polypropylene reinforced with nano zinc oxide. The results show that the Melt temperature (Tm) and crystallization temperature (Tc) did not change very much with the addition of ZnO (Table 11).

Table 11. Thermal properties of the composites

Blend	T_m (°C)	T_c (°C)	Enthalpy (J/g)	X_c(%)
Original PP	166,3	124,9	85,4	40,86
PP/1ZnO	166,6	123,1	65	31,10
PP/3ZnO	165,9	123,4	63,42	30,34
PP/5ZnO	165,9	123,1	64,6	30,91

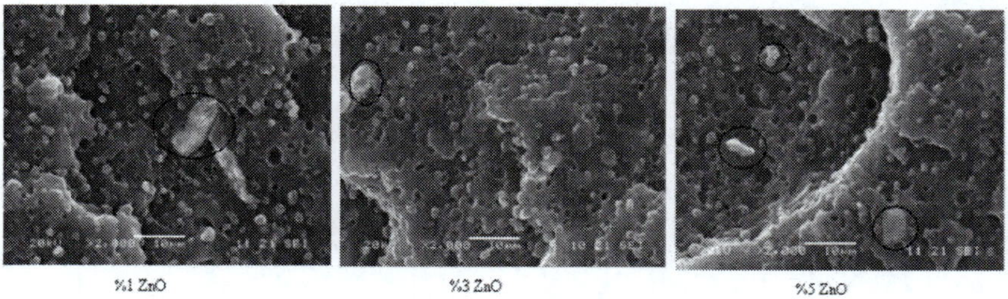

Figure 31. SEM images of PP/ZnO nanocomposites.

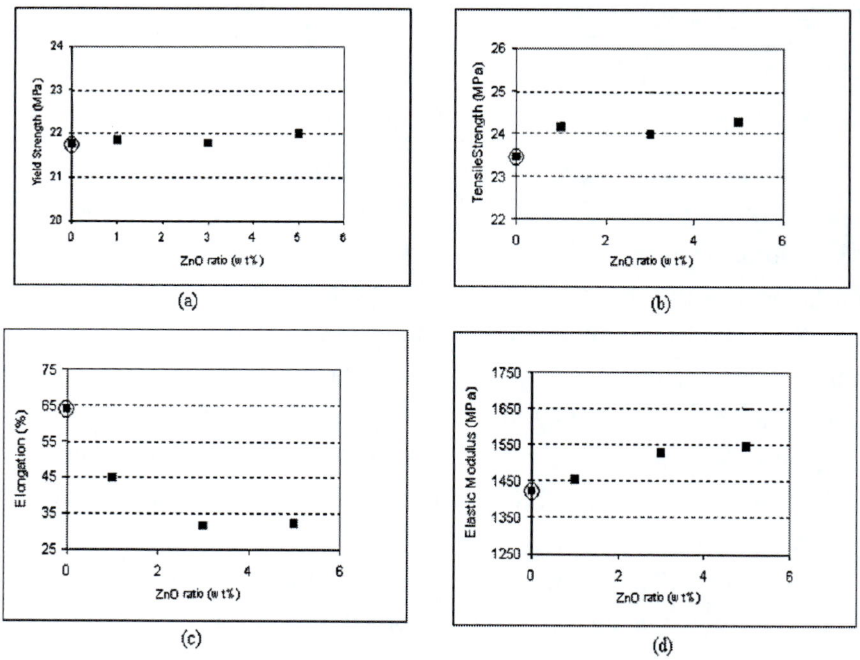

Figure 32. Tensile properties of nano ZnO reinforced PP (a-yield strength, b-tensile modulus, c- elongation d-elastic modulus) and neat PP.

The possible agglomerates given in Figure 31 within black circles strengthen the lower crystallinity of the ZnO composites. According to Figure 32, the tensile properties such as tensile strength, elastic modulus, and stress at break increase while elongation decreases.

Polypropelene/Graphene Nanocomposite

The first study on graphene, or two-dimensional graphite, can be dated as early as 1947 when Wallace used the 'tight binding' approximation to investigate the electronic energy bands in crystalline graphite. Due to the internal exceptionally high crystal quality and massless Dirac fermions, monolayer graphene exhibits anomalous half-integer quantum Hall effect, remarkable optical properties, ultra-high intrinsic strength, superior thermal conductivity, and extremely high charge carrier mobility. Graphene-based materials are expected to be promising building blocks in nanotechnology due to their varieties of applications, as seen in Figure 33 [73].

One of the primary concerns in graphene synthesis is producing samples with high carrier mobility and low density of defects. To date, there is no method that can match mechanical exfoliation for producing high-quality and high mobility graphene flakes. However, mechanical exfoliation is a time-consuming process limited to small-scale production. There is great interest in producing large-scale graphene suitable for applications in flexible transparent electronics, transistors, and so forth. Some concerns in producing large-scale graphene are the quality and consistency between samples as well as the cost and difficulty involved in the method. Table 12 shows a summary of some of the most important synthesis methods. Oxidation of graphite to graphene oxide (GO) using various oxidants has been successfully employed as a precursor for graphene synthesis. The graphene structure has been investigated through numerous theoretical and experimental routes. Owing to the excellent electrical, mechanical, and structural properties, graphene has been focused on the fabrication of polymer nanocomposites. Graphene has become preferred and significantly better nanofiller for polymers than other nanocarbon and inorganic nanofiller materials [73, 74].

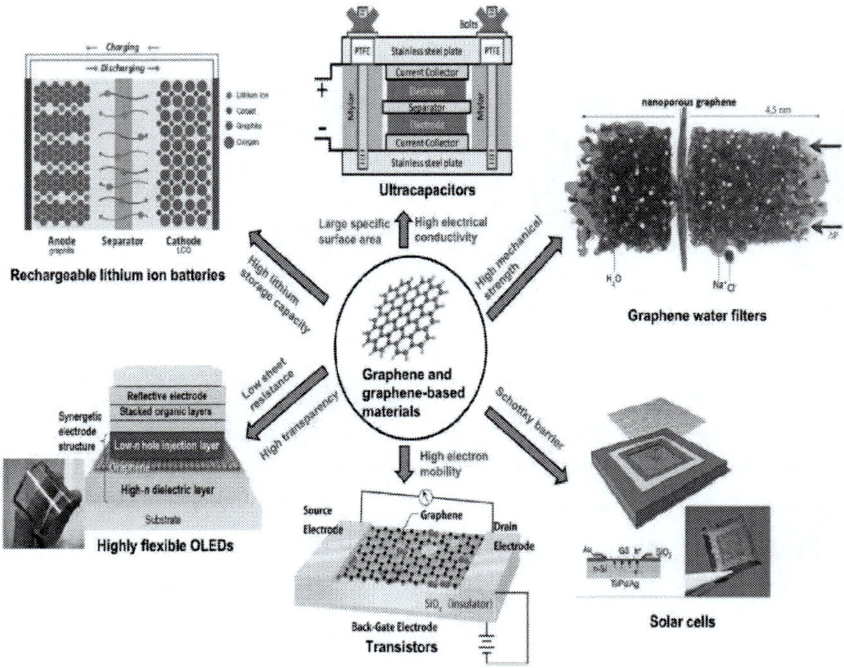

Figure 33. Applications of graphene and graphene-based materials in batteries, ultracapacitor, water filters, solar cells, transistors, and OLEDs.

Table 12. Comparison of graphene synthesis methods. Shows the typical number of layers produced, size of graphene layers

Method	Layers	Size	Mobility ($cm^2V^{-1}s^{-1}$)
Exfoliation	1 to 10+	1 mm[a]	15000[b]
Thermal SiC	1 to 4	50 μm[c]	2000[c]
Ni-CVD	1 to 4	1 cm[d]	3700[d]
Cu-CVD	1	65 cm[e]	16000[f]

Due to the promising extraordinary functional properties, graphene has been used an efficient nanofiller for polymeric nanocomposite

(Figure 34). The polymer/graphene nanocomposite has been an exciting area of research for many years, and several polymers have been amalgamated with graphene [74].

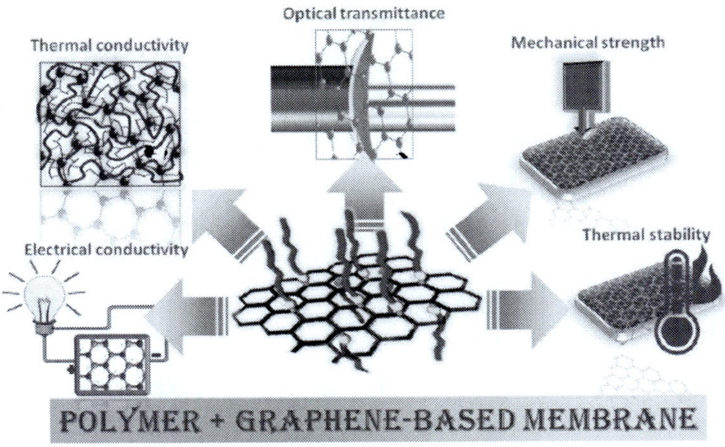

Figure 34. Graphene nanofiller affecting polymeric membrane properties.

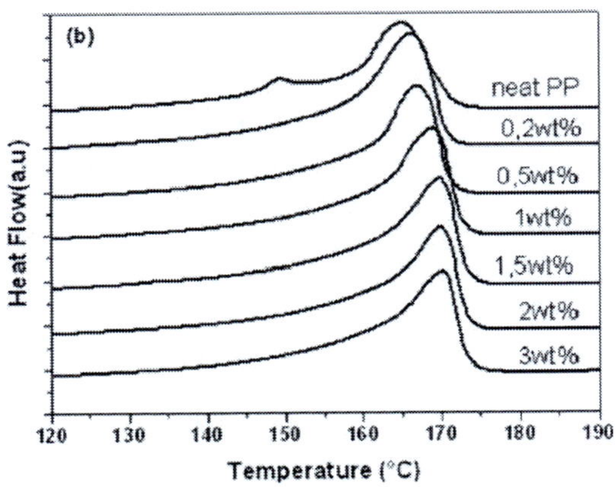

Figure 35. Crystallization thermograms of neat PP and its nanocomposites with various GNs contents of 0.2, 0.5, 1.0, 1.5, 2.0, and 3.0 wt%.

Achaby et al. [75] were study the mechanical, thermal, and rheological properties of graphene-based polypropylene nanocomposites prepared by melt mixing. The results of the differential scanning calorimetry test show two important differences between neat PP and its nocomposites (Figure 34). First, the neat PP shows two main melting peaks, while all the nanocomposites containing GNs exhibit only one peak. The second difference is the increase in the melting temperature with increasing GNs content. For neat PP, the peak at 164.98C is attributed to the melting of α-crystals, the most common monoclinic crystal structure of PP, whereas the peak at 147.58C is attributed to the melting of β-crystals.

In addition, the thermal degradation of PP and PP nanocomposites with different weight fractions of GNs was determined from the weight loss during heating. Thermal degradation of the neat PP and its nanocomposites occurs as a single step process, with a maximum decomposition temperature at around 460°C, while the values at a weight loss of 5 and 15% for neat PP are at 396 and 428°C, respectively. The addition of GNs increases the thermal stability by around 408°C for T5%, 208°C for T15%, and 108°C for T_{max}. The improvement in the resistance to thermal degradation can be attributed to the hindered diffusion of volatile decomposition products within the nanocomposites. It is strongly dependent on the nanosheets polymer chains interactions.

Table 13. Summarized results of TGA analysis (T5%, T15%, and T_{max}) of PP and its nanocomposites with different GNs content

GNs content		TGA		DTG
Wt%	Vol%	$T_{5\%}$ (°C)	$T_{15\%}$ (°C)	T_{max} (°C)
0	—	396.15	428.50	460.13
0.2	0.08	405.32	431.73	461.53
0.5	0.2	407.34	434.82	462.15
1	0.4	415.14	437.67	465.23
1.5	0.6	424.83	443.11	467.98
2	0.8	433.64	446.52	469.37
3	1.2	436.57	450.00	470.18

The results show significant improvements in mechanical properties by incorporating GNs into the PP matrix. In fact, typical stress-strain curves obtained by uniaxial tensile testing of the PP matrix and its GNs nanocomposites are presented in Figure 35a. The extracted values of the

Young modulus, tensile strength, and elongation at break are summarized in Table 14 and they are plotted in Figure 35b and c as a function of GNs content. The results show that the Young modulus and tensile strength increase significantly with increasing GNs content due to their high mechanical strength. The addition of only 1 wt% (0.4 vol%) of GNs results in an increase of the Young modulus and the tensile strength of about 64 and 54%, respectively, as compared with neat PP values. As the loading of GNs increases to 3 wt% (1.2 vol%), the Young modulus and tensile strength of PP/GNs nanocomposites show a substantial increase of about 100 and 81%, respectively. As can be expected for any filled polymer, the addition of GNs.

Table 14. Summarized results of mechanical properties and its % increase of neat PP and its nanocomposites with different GNs content

GNs content		Young's Modulus, E (Mpa)	% increase of E	Tensile strength, σ (Mpa)	% increase of σ	Elongation at break (%)
Wt%	Vol%					
0	—	1154.48	—	33.98	—	279.91
0.2	0.08	1252.21	14.66	37.58	10.59	36.33
0.5	0.2	1576.96	36.59	38.81	14.21	28.34
1	0.4	1785.33	54.64	43.93	29.28	27.49
1.5	0.6	1888.27	63.55	51.37	51.17	26.28
2	0.8	2040.04	76.70	56.17	65.30	21.71
3	1.2	2314.61	100.48	61.57	81.19	19.09

Mechanical properties, including hardness, elastic modulus, and fracture toughness of the prepared GO/PP nanocomposites were analyzed by Shin et al. [76]. The fracture toughness (Kc) was calculated from the hardness, elastic modulus, and crack size after nanoindentation. Based on a half - penny crack configuration, Lawn et al. proposed the following expression for fracture toughness.

$$K_c = (\frac{E}{H})^2 \frac{P}{c^{\frac{3}{2}}} \quad (19)$$

where k is a calibration constant (0.016), and the value of the exponent n was estimated to be $\frac{1}{2}$. For center-loaded cracks, $\frac{P}{c^{\frac{3}{2}}}$ is a constant, c = l + a, and $(\frac{a}{l}) \cong 0.68$. The crack length l in m can be measured from the impression corner. To gain insight into crack propagation behavior of the GO/PP nanocomposites, the fracture energy (G_c) of GO/PP nanocomposites was measured based on the elastic modulus and fracture toughness as follows:

$$G_c = K_c^2 \left(\frac{1-\nu^2}{E}\right) \quad (20)$$

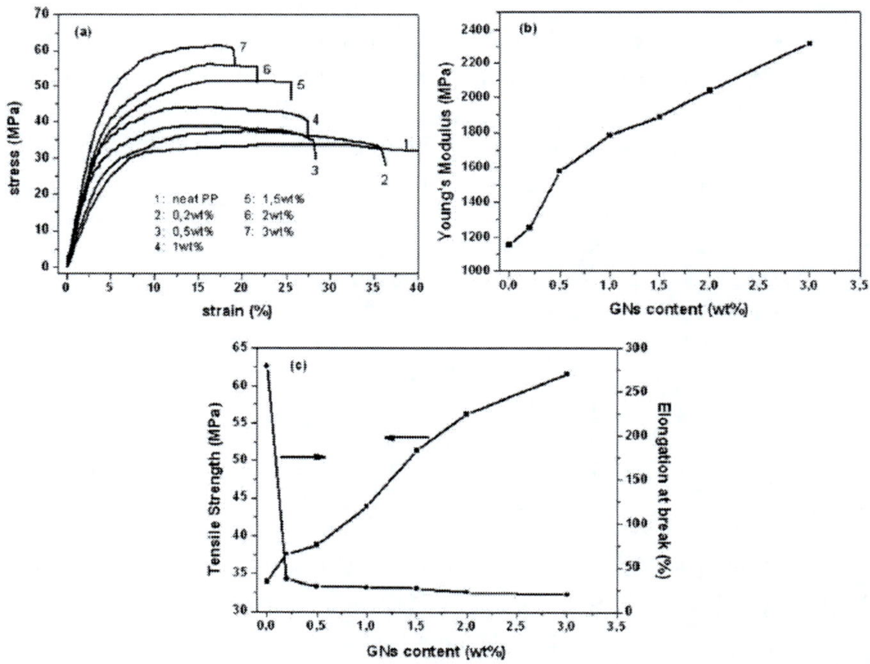

Figure 36. (a) Typical stress-strain curves of the neat PP and its nanocomposites at indicated GNs contents (crosshead speed set at 10 mm/ min), (b) Young's modulus, and (c) tensile strength and elongation at break of GNs-filled PP nanocomposites as a function of GNs content.

The fracture energy is the energy required for a crack to propagate in the GO/PP nanocomposites. The mechanical properties of several GO/PP nanocomposites are listed in Table 15.

Table 15. Mechanical properties of prepared GO/PP nanocomposites

Samples	Hardness[a]/MPa	Elastic modulus[a]/GPa	Fracture toughness[a]/MPa $m^{1/2}$	Fracture energy[a]/kJ m^{-2}	Surface roughness[a]/nm
1	6.616	0.459	1.654	5.424	3.521
2	7.156	0.539	1.862	5.853	8.406
3	8.028	0.592	2.263	8.583	9.498
4	8.669	0.666	2.205	8.656	12.997
5	21.793	0.613	2.694	10.774	5.444
6	25.451	0.651	2.721	10.350	6.016
7	25.562	0.743	3.589	11.238	9.136

[a] All the properties were measured with a Digital Instrument Nanoscope IIIA from Veeco Systems in tapping mode using silicon tips with a resonance frequency of 320 kHz and a trigger threshold of 5.0 V. The elastic modulus and Poisson's ratio for the nanoindenter were 0.4098 and 0.2780, respectively, and the Poisson's ratio of GO/PP nanocomposites was assumed to be *ca.* 0.3.

Scratch hardness (H_s) was evaluated to characterize the resistance to scratch damage and was determined using the following relationship, which is analogous to nanoindentation hardness:

$$H_s = \frac{N_s}{A_p} \text{ For a conical probe } A_P = \frac{\pi w^2}{4} \tag{21}$$

$$H_s = \frac{4xN_s}{\pi w^2} \text{ X is 2 for rigid plastic materials and } 1 < x < 2 \text{ for viscoelastic material} \tag{22}$$

where H_s is the scratch hardness in $\frac{kgf}{mm^2}$, N_s is the scratch probe load in Newtons, and A_P is the sliding contact area of the scratch probe load in mm^2.

The relationship between the scratch width and scratch depth (d) for a conical probe with an angle of 120° is $w = 2\sqrt{3}d$. Therefore, the scratch hardness can be estimated based on the scratch depth by:

$$H_s = \frac{N_s}{3\pi d^2} \tag{23}$$

The scratch hardness of several GO/PP nanocomposites with respect to the probe width, load, and scratch speed is shown in Table 2.

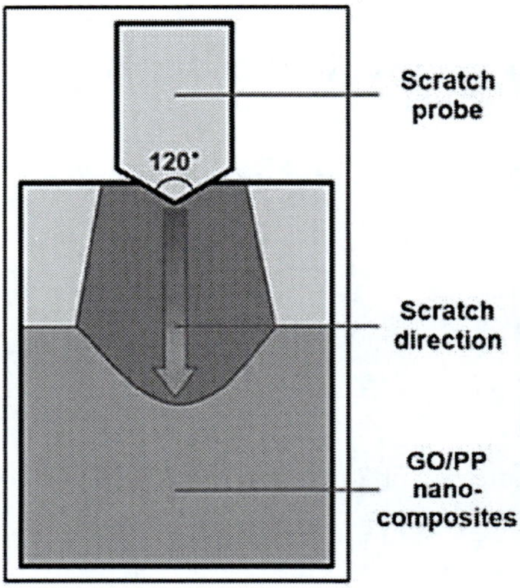

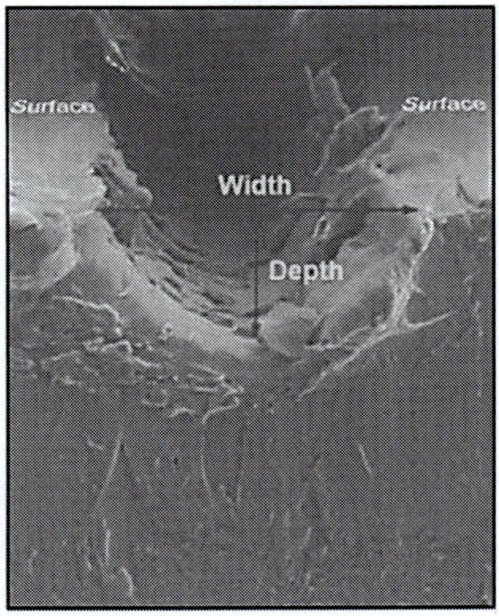

Figure 36. The schematic diagram of the scratch test and SEM micrograph showing the cross-section of scratched GO/PP nanocomposites.

Table 16. Scratch hardness of GO/PP nanocomposites by controlling the scratch parameter

	Scratch hardness (parameter: probe width, μm per load, kgf per scratch speed, m min^{-1})[a]						
Samples	100/0.5/10	200/0.5/10	500/0.5/10	500/0.5/3	500/0.5/20	500/0.1/10	500/0.3/10
1	34.1	97.5	46.1	38.9	62.1	13.9	41.6
2	54.3	120.0	69.3	56.6	82.5	24.0	44.8
3	61.7	132.5	94.5	84.3	132.5	38.9	52.1
4	71.2	136.2	97.5	86.9	142.0	61.9	56.7
5	57.3	136.2	82.5	60.5	136.2	48.1	48.9
6	76.8	162.4	125.9	108.3	141.5	117.6	119.2
7	82.5	174.7	149.9	131.1	223.7	129.1	144.4

[a] The scratch hardness was calculated by the average of scratch hardness value obtained by the scratch width and depth, and the viscoelasticity of all samples was assumed to be 1. The scratch width and depth of each sample were measured by the SEM analysis.

The results show that 286% increase in hardness, 127% improvement of elastic modulus, and 117% enhancement on fracture toughness are achieved by adding only 1.0 wt% of GO.

REFERENCES

[1] Nakhaei MR, Mostafapour A, Dubois C, Naderi G, Reza Ghoreishy MH (2019) Study of morphology and mechanical properties of PP/EPDM/clay nanocomposites prepared using twin-screw extruder and friction stir process. *Polymer Composites* 40: 3306-3314.

[2] Haghnegahdar M, Naderi G, Ghoreishy M (2017) Fracture toughness and deformation mechanism of un-vulcanized and dynamically vulcanized polypropylene/ethylene propylene diene monomer/graphene nanocomposites. *Composites Science and Technology* 141: 83-98.

[3] Zou W, Chen R-y, Wu C, Qu J-p (2016) Influence of process parameters on property of PP/EPDM blends prepared by a novel vane extruder. *Journal of Polymer Engineering* 36: 899-908.

[4] Mostafapour A, Naderi G, Nakhaei MR (2018) Effect of process parameters on fracture toughness of PP/EPDM/nanoclay nanocomposite fabricated by novel method of heat assisted Friction stir processing. *Polymer Composites* 39: 2336-2346.

[5] Ahmadi A, Arab NM, Naderi G, Nakhaei M (2017) Optimization of CO2 laser welding process parameters of PP/EPDM/Clay nanocomposite using response surface methodology. *Mechanics & Industry* 18: 220.

[6] Bhattacharya M (2016) Polymer nanocomposites—a comparison between carbon nanotubes, graphene, and clay as nanofillers. *Materials* 9: 262.

[7] Hashim A (2011). *Advances in nanocomposite technology.*

[8] Boussia AC, Vouyiouka SN, Papaspyrides CD (2011) Polyamide Nanocomposites by In-situ Polymerization. *In-Situ Synthesis of Polymer Nanocomposites*: 27-51.

[9] Hu S, Zhou Y, Yuan C, Wang W, Hu J, Li Q, He J (2020) Surface-modification effect of MgO nanoparticles on the electrical properties of polypropylene nanocomposite. *High Voltage* 5: 249-255.

[10] Cabello-Alvarado C, Reyes-Rodríguez P, Andrade-Guel M, Cadenas-Pliego G, Pérez-Alvarez M, Cruz-Delgado V, Melo-López L, Quiñones-Jurado Z, Ávila-Orta C (2019) Melt-mixed thermoplastic nanocomposite containing carbon nanotubes and titanium dioxide for flame retardancy applications. *Polymers* 11: 1204.

[11] Khare V, Kamle S (2020) Optimization and assessment of MWCNT dispersion in polypropylene nanocomposite films. *Journal of Thermoplastic Composite Materials*: 08927057 20962169.

[12] Safaie B, Youssefi M, Rezaei B (2018) Estimating the interphase properties of polypropylene/carbon quantum dot nanocomposite fibers by micromechanical modeling. *Colloid and Polymer Science* 296: 1953-1960.

[13] Haghnegahdar M, Naderi G, Ghoreishy M (2017) Microstructure and mechanical properties of nanocomposite based on polypropylene/ethylene propylene diene monomer/graphene. *International Polymer Processing* 32: 72-83.

[14] Tripathi SN, Rao GS, Mathur AB, Jasra R (2017) Polyolefin/ graphene nanocomposites: a review. *RSC advances* 7: 23615-23632.

[15] Giannakas A, Patsaoura A, Barkoula N-M, Ladavos A (2017) A novel solution blending method for using olive oil and corn oil as plasticizers in chitosan based organoclay nanocomposites. *Carbohydrate polymers* 157: 550-557.

[16] Tommasini FJ, Ferreira LdC, Tienne LGP, Aguiar VdO, Silva MHPd, Rocha LFdM, Marques MdFV (2018) Poly (methyl methacrylate)-SiC nanocomposites prepared through in situ polymerization. *Materials Research* 21.

[17] Sun H, Wang T, Xu Y, Gao W, Li P, Niu QJ (2017) Fabrication of polyimide and functionalized multi-walled carbon nanotubes mixed matrix membranes by in-situ polymerization for CO2 separation. *Separation and Purification Technology* 177: 327-336.

[18] Shen H, Wang S, Xu H, Zhou Y, Gao C (2018) Preparation of polyamide thin film nanocomposite membranes containing silica nanoparticles via an in-situ polymerization of SiCl4 in organic solution. *Journal of Membrane Science* 565: 145-156.

[19] Sheng X, Zhao Y, Zhang L, Lu X (2019) Properties of two-dimensional Ti3C2 MXene/thermoplastic polyurethane nanocomposites with effective reinforcement via melt blending. *Composites Science and Technology* 181: 107710.

[20] Ercan N, Durmus A, Kaşgöz A (2017) Comparing of melt blending and solution mixing methods on the physical properties of thermoplastic polyurethane/organoclay nanocomposite films. *Journal of Thermoplastic Composite Materials* 30: 950-970.

[21] Chouikhi N, Cecilia JA, Vilarrasa-García E, Besghaier S, Chlendi M, Franco Duro FI, Rodriguez Castellon E, Bagane M (2019) CO2 adsorption of materials synthesized from clay minerals: A review. *Minerals* 9: 514.

[22] Lu Y, Dong W, Wang W, Ding J, Wang Q, Hui A, Wang A (2018) Optimal synthesis of environment-friendly iron red pigment from natural nanostructured clay minerals. *Nanomaterials* 8: 925.

[23] Liu P (2007) Polymer modified clay minerals: A review. *Applied Clay Science* 38: 64-76.

[24] Zabihi O, Ahmadi M, Nikafshar S, Preyeswary KC, Naebe M (2018) A technical review on epoxy-clay nanocomposites: Structure, properties, and their applications in fiber reinforced composites. *Composites part b: engineering* 135: 1-24.

[25] Ruiz-Hitzky E, Van Meerbeek A (2006) Clay mineral–and organoclay–polymer nanocomposite. *Developments in Clay Science* 1: 583-621.

[26] Mishra R, Manral A (2021) Graphene Functionalized Starch Biopolymer Nanocomposites: Fabrication, Characterization, and Applications. In *Graphene Based Biopolymer Nanocomposites*, Ed., Springer. 173-189.

[27] Koo JH (2019). *Polymer nanocomposites: processing, characterization, and applications* McGraw-Hill Education.

[28] Sharma B, Jain P (2021). *Graphene Based Biopolymer Nanocomposites* Springer.

[29] Saravanan N, Rajasekar R, Mahalakshmi S, Sathishkumar T, Sasikumar K, Sahoo S (2014) Graphene and modified graphene-based polymer nanocomposites–a review. *Journal of Reinforced Plastics and Composites* 33: 1158-1170.

[30] Petkov P, Achour ME, Popov C (2020). *Nanoscience and Nanotechnology in Security and Protection against CBRN Threats* Springer Nature.

[31] Gdoutos EE (2020). *Fracture mechanics: an introduction*. Vol. 263. Springer Nature.

[32] Knott JF (1973). *Fundamentals of fracture mechanics* Gruppo Italiano Frattura.

[33] Kinloch AJ (2013). *Fracture behaviour of polymers* Springer Science & Business Media.

[34] Kausch HH (2012). *Polymer fracture. Vol. 2*. Springer Science & Business Media.

[35] Lu ML, Chiou KC, Chang FC (1996) Elastic-plastic fracture toughness of PC/ABS blend based on CTOD and J-integral methods. *Polymer* 37: 4289-4297.

[36] Moore DR, Williams J, Pavan A (2001). *Fracture mechanics testing methods for polymers, adhesives and composites* Elsevier.

[37] Avci A, Arikan H, Akdemir A (2004) Fracture behavior of glass fiber reinforced polymer composite. *Cement and Concrete Research* 34: 429-434.

[38] Lu ML, Lee CB, Chang FC (1995) Fracture toughness of acrylonitrile-butadiene-styrene by J-integral methods. *Polymer Engineering & Science* 35: 1433-1439.

[39] Narisawa I, Takemori M (1989) Fracture toughness of impact-modified polymers based on the J-integral. *Polymer Engineering & Science* 29: 671-678.

[40] Grellmann W, Seidler S, Anderson PI (2007). *Polymer testing Hanser Munich.*

[41] Bhurke AS, Shin EE, Drzal LT (1997) Fracture morphology and fracture toughness measurement of polymer-modified asphalt concrete. *Transportation Research Record* 1590: 23-33.

[42] Li Y, Wang S, Wang Q, Xing M (2018) Enhancement of fracture properties of polymer composites reinforced by carbon nanotubes: a molecular dynamics study. *Carbon* 129: 504-509.

[43] Agnelli S, Ramorino G, Passera S, Karger-Kocsis J, Ricco T (2012) Fracture resistance of rubbers with MWCNT, organoclay, silica and carbon black fillers as assessed by the J-integral: Effects of rubber type and filler concentration. *Express Polymer Letters* 6.

[44] Grellmann W, Lach R, Seidler S (2002) Experimental determination of geometry-independent fracture mechanics values J, CTOD and K for polymers. *International journal of fracture* 118: 9-14.

[45] Khodabandelou M, Aghjeh MKR, Mazidi MM (2015) Fracture toughness and failure mechanisms in un-vulcanized and dynamically vulcanized PP/EPDM/MWCNT blend-nanocomposites. *RSC Advances* 5: 70817-70831.

[46] Paran S, Naderi G, Ghoreishy M, Dubois C (2018) Essential work of fracture and failure mechanisms in dynamically vulcanized thermoplastic elastomer nanocomposites based on PA6/NBR/XNBR-grafted HNTs. *Engineering Fracture Mechanics* 200: 251-262.

[47] Poon W, Ching E, Cheng C, Li R (2001) Measurement of plane stress essential work of fracture (EWF) for polymer films: effects of gripping and notching methodology. *Polymer testing* 20: 395-401.

[48] Gomari S, Ghasemi I, Karrabi M, Azizi H (2015) *An investigation on non-isothermal crystallization behavior and morphology of polyamide 6/poly (ethyleneco-1-butene)-graft-maleic anhydride/organoclay nanocomposites.*

[49] nakhaei m, Bani Mostafa Arab N, rezaei g (2015) Investigation on Tensile Strength of Friction Stir Welded Joints in PP/EPDM/Clay Nanocomposites. *International Journal of Engineering* 28: 1382-1391.

[50] Taghizadeh E, Naderi G, Razavi-Nouri M (2011) Effects of organoclay on the mechanical properties and microstructure of PA6/ECO blend. *Polymer Testing* 30: 327-334.

[51] Nam PH, Maiti P, Okamoto M, Kotaka T, Hasegawa N, Usuki A (2001) A hierarchical structure and properties of intercalated polypropylene/clay nanocomposites. *Polymer* 42: 9633-9640.

[52] Zheng W, Lu X, Ling Toh C, Hua Zheng T, He C (2004) Effects of clay on polymorphism of polypropylene in polypropylene/clay nanocomposites. *Journal of Polymer Science Part B: Polymer Physics* 42: 1810-1816.

[53] Mohan T, Kanny K (2011) Influence of nanoclay on rheological and mechanical properties of short glass fiber-reinforced polypropylene composites. *Journal of Reinforced Plastics and Composites* 30: 152-160.

[54] Maiti P, Nam PH, Okamoto M, Hasegawa N, Usuki A (2002) Influence of crystallization on intercalation, morphology, and mechanical properties of polypropylene/clay nanocomposites. *Macromolecules* 35: 2042-2049.

[55] Kodgire P, Kalgaonkar R, Hambir S, Bulakh N, Jog J (2001) PP/clay nanocomposites: effect of clay treatment on morphology and dynamic mechanical properties. *Journal of Applied Polymer Science* 81: 1786-1792.

[56] Rajesh J, Soulestin J, Lacrampe MF, Krawczak P (2012) Effect of injection molding parameters on nanofillers dispersion in masterbatch based PP-clay nanocomposites. *EXPRESS polymer Letters* 6: 237-248.

[57] Zhang M, Li J (2009) Carbon nanotube in different shapes. *Materials today* 12: 12-18.

[58] Ma PC, Siddiqui NA, Marom G, Kim JK (2010) Dispersion and functionalization of carbon nanotubes for polymer-based nanocomposites: A review. *Composites Part A: Applied Science and Manufacturing* 41: 1345-1367.

[59] Sharma R, Sharma AK, Sharma V (2015) Synthesis of carbon nanotubes by arc-discharge and chemical vapor deposition method

with analysis of its morphology, dispersion and functionalization characteristics. *Cogent Engineering* 2: 1094017.

[60] Jeon IY, Chang DW, Kumar NA, Baek JB (2011) Functionalization of carbon nanotubes. *Carbon nanotubes-polymer nanocomposites:* 91-110.

[61] Kocabas C, Kang SJ, Ozel T, Shim M, Rogers JA (2007) Improved synthesis of aligned arrays of single-walled carbon nanotubes and their implementation in thin film type transistors. *The Journal of Physical Chemistry* C 111: 17879-17886.

[62] Esmaeili M, Ansari R, Mottaghitalab V, Haghi A, Xi B, Spinks GM, Wallace GG, Minett AI (2014) CNT/polymer nanocomposites: a detailed review on mathematical modeling and experimental case studies. Engineering of Polymers and Chemical Complexity: *Current State of the Art and Perspectives* 1: 149.

[63] Bao S, Tjong SC (2008) Mechanical behaviors of polypropylene/carbon nanotube nanocomposites: The effects of loading rate and temperature. *Materials Science and Engineering: A* 485: 508-516.

[64] Yetgin SH (2019) Effect of multi walled carbon nanotube on mechanical, thermal and rheological properties of polypropylene. *Journal of Materials Research and Technology* 8: 4725-4735.

[65] Wang P-H, Sarkar S, Gulgunje P, Verghese N, Kumar S (2018) Fracture mechanism of high impact strength polypropylene containing carbon nanotubes. *Polymer* 151: 287-298.

[66] Wacharawichanant S, Thongyai S, Siripattanasak T, Tipsri T (2009) *Effect of mixing conditions and particle sizes of titanium dioxide on mechanical and morphological properties of polypropylene/titanium dioxide composites.*

[67] Aydemir D, Gulsen U, GUMUŞ H, Yildiz S, GUMUŞ S, Bardak T, Gunduz G (2016) Nanocomposites of polypropylene/nano titanium dioxide: effect of loading rates of nano titanium dioxide. *Materials Science/Medžiagotyra* 22: 364-369.

[68] Esthappan SK, Kuttappan SK, Joseph R (2012) Thermal and mechanical properties of polypropylene/titanium dioxide nanocomposite fibers. *Materials & Design* 37: 537-542.

[69] Chandramouleeswaran S, Mhaske S, Kathe A, Varadarajan P, Prasad V, Vigneshwaran N (2007) Functional behaviour of polypropylene/ZnO–soluble starch nanocomposites. *Nanotechnology* 18: 385702.

[70] Bustos-Torres KA, Vazquez-Rodriguez S, Martínez-de la Cruz A, Sepulveda-Guzman S, Benavides R, Lopez-Gonzalez R, Torres-Martínez LM (2017) Influence of the morphology of ZnO nanomaterials on photooxidation of polypropylene/ZnO composites. *Materials Science in Semiconductor Processing* 68: 217-225.

[71] Karami Z, Youssefi M, Borhani S (2013) The effects of UV irradiation exposure on the structure and properties of polypropylene/ZnO nanocamposite fibers. *Fibers and Polymers* 14: 1627-1634.

[72] Altan M, Yildirim H, Uysal A (2011) Tensile properties of polypropylene/metal oxide nano composites. *The Online Journal of Science and Technology* 1: 25-30.

[73] Singh V, Joung D, Zhai L, Das S, Khondaker SI, Seal S (2011) Graphene based materials: past, present and future. *Progress in materials science* 56: 1178-1271.

[74] Jo JW, Lee JU, Jo WH (2015) Graphene-based electrodes for flexible electronics. *Polymer International* 64: 1676-1684.

[75] El Achaby M, Arrakhiz FE, Vaudreuil S, el Kacem Qaiss A, Bousmina M, Fassi-Fehri O (2012) Mechanical, thermal, and rheological properties of graphene-based polypropylene nanocomposites prepared by melt mixing. *Polymer composites* 33: 733-744.

[76] Shin KY, Hong JY, Lee S, Jang J (2012) Evaluation of anti-scratch properties of graphene oxide/polypropylene nanocomposites. *Journal of Materials Chemistry* 22: 7871-7879.

INDEX

#

A

B

C

D

O

P

R

S

T

W

X

Z